• THE R.C. RILEY COLOUR COLLECTION •

MIDLAND REGION

3RD JUNE 1962 • 46200 by the coaling stage at Camden. RCR5556

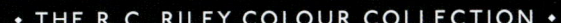

IMAGES FROM THE TRANSPORT TREASURY ARCHIVE • COMPILED BY PETER SIKES

THE R.C. RILEY COLOUR COLLECTION • MIDLAND REGION

3RD JUNE 1962 • CORONATION at Camden. RCR5588

Images and design © The Transport Treasury 2024. Design and Text: Peter Sikes
ISBN: 978-1-913251-83-3
First published in 2024 by Transport Treasury Publishing Ltd., 16 Highworth Close, High Wycombe HP13 7PJ

The copyright holders hereby give notice that all rights to this work are reserved.
Aside from brief passages for the purpose of review, no part of this work may be reproduced, copied by electronic or other means,
or otherwise stored in any information storage and retrieval system without written permission from the Publisher.
This includes the illustrations herein which shall remain the copyright of the copyright holder.
Copies of many of the images in THE R.C. RILEY COLOUR COLLECTION – MIDLAND REGION
are available for purchase/download, the reference codes are at the end of the captions.
In addition the Transport Treasury Archive contains tens of thousands of other UK, Irish and some European railway photographs.

www.ttpublishing.co.uk

Printed in England by Short Run Press Limited, Exeter.

INTRODUCTION

Railways, particularly steam, and photography were very much part of Riley family life. Richard had this enormous enthusiasm and expertise which resulted in a huge archive of colour slides, which have continued to be published since his death in 2006. His passion for this subject was never in doubt, but he was always modest about his place in being one of the country's most talented and prolific railway photographers.

Often described as a gentleman, Richard was also a gentle man which often shone through when he was approached by other aspiring photographers and authors. He willingly gave his time to help others in many different fields. Equally, his family was very important to him. If ever advice and support were needed by the immediate family, his wisdom and integrity could always be relied upon, and he is much missed by us all.

Christine Riley

25TH MAY 1963 • Rebuilt 'Royal Scot' No. 46140 THE KING'S ROYAL RIFLE CORPS pictured passing Willesden. RCR5596

THE R.C. RILEY COLOUR COLLECTION • MIDLAND REGION

3RD JUNE 1962 • No. 46200 THE PRINCESS ROYAL heads towards Euston station as it passes through Camden with the Up 'Aberdeen Flyer'. RCR5553

THE ABERDEEN FLYER RAIL TOUR – 1X76 – 2nd/3rd June 1962 The Railway Correspondence and Travel Society/ Stephenson Locomotive Society	
60022	London Kings Cross – Selby – York – Newcastle – Edinburgh Waverley
60004	Edinburgh Waverley – Dundee Tay Bridge – Aberdeen
GNSR 49	Waterloo Goods – Kittybrewster – Dyce – Inverurie (Works Siding)
GNSR 49	Inverurie (Works Siding) – Old Meldrum
65345	Old Meldrum – Inverurie
65345	Inverurie – Dyce – Kittybrewster – Aberdeen
46201	Aberdeen – Kinnaber Jn – Bridge of Dun – Forfar – Perth – Glenboig – Coatbridge – Motherwell – (via WCML) – Carlisle
46200	Carlisle – (via WCML) – Crewe
46200	Crewe – (via WCML) – London Euston

The R.C.T.S. and Stephenson Locomotive Society organised a rail tour over the weekend of 2nd and 3rd June 1962. The 'Aberdeen Flyer' departed Kings Cross behind A4 No. 60022 MALLARD. No. 60004 WILLIAM WHITELAW was in charge of the train from Edinburgh Waverley to Aberdeen, GNSR No. 49 GORDON HIGHLANDER and J36 0-6-0 No. 65345 then took the train to Inverurie (Works Siding), Old Meldrum and back to Aberdeen via Inverurie and Kittybrewster. 'Princess Royal' Class Pacific No. 46201 PRINCESS ELIZABETH took over at Aberdeen and ran via Perth and Motherwell to the West Coast Main Line and onwards to Carlisle. There No. 46200 THE PRINCESS ROYAL hauled the final legs (1X76) from Carlisle to Crewe, then Crewe to London Euston.

RIGHT: No. 46200 is pictured at Camden shed after replenishing coal and water supplies, ready to head back north. At the time the loco was allocated to Carlisle Kingmoor (12A). RCR5561

When William A. Stanier became CME of the LMS in 1932 he wasted no time in setting about dealing with the company's motive power problems. Within 17 months of taking office, the first Stanier designed locomotive entered traffic on 1st July 1933. This was No. 6200, the first of the 'Princess Royal' class Pacifics for express passenger work. Introduced for Anglo-Scottish services, the two prototypes included domeless taper Belpaire boilers with low temperature superheat and top feed. It was this small superheat which was to prove troublesome under LMS operating conditions. A further batch of ten engines (Nos. 6203-6212) were built at Crewe in 1935; these were built with a modified boiler with shorter tubes and a larger firebox. They became successful locomotives but were gradually overshadowed by the 'Princess Coronations' although, like the later Pacifics, they were a picture of elegance and power.

ABOVE: 3RD JUNE 1962 • A final view of No. 46200 by the coaling stage at Camden accompanied by an unidentified 'Royal Scot'. RCR5543

OPPOSITE PAGE

TOP: 16TH FEBRUARY 1958 • No prizes for guessing where Mr. Riley was on this date. RCR0469

MIDDLE: 20TH SEPTEMBER 1958 • Caprotti 'Black Five' No. 44741, at this time allocated to Longsight (9A), in the company of an unidentified conventional class member near to the coaling plant at Camden. RCR5687

BOTTOM: 3RD JUNE 1962 • The first of Stanier's impressive Class 8P 'Princess Coronation' Pacifics, No. 46220 CORONATION, pictured in light steam at Camden. RCR5555

CAMDEN MOTIVE POWER DEPOT

The original Camden shed was built by the London and Birmingham Railway in 1837 and was located at nearby Chalk Farm. Due to increased traffic new locomotive facilities were required. Two buildings were constructed on either side of the main line into Euston, one for freight locomotives (the Roundhouse) and a larger depot for passenger locomotives. The roundhouse was closed in 1871 and replaced by Willesden motive power depot, the building being converted into a storage warehouse.

The passenger depot was enlarged by the London and North Western Railway in 1920 and then in 1932, as far as the cramped location would allow, the London, Midland and Scottish Railway altered and modernised the site. This included the addition in 1936 of modern ash and coaling plants and a new 70 foot vacuum operated turntable. A wheel drop was installed in the former fitting shop and a machine shop provided. Camden shed was always a cramped and busy place and there were several notices around the site warning engine crew not to generate black smoke due to its proximity to housing and flats.

The depot was closed to steam locomotives by British Railways in September 1963 and briefly used as a diesel depot until 1966, when it was demolished and replaced by sidings.

ABOVE: 21ST AUGUST 1955 • The driver of rebuilt 'Royal Scot' No. 46148 THE MANCHESTER REGIMENT oils his loco at Camden shed. No. 46148 was renamed in October 1935, previously carrying the name VELOCIPEDE from introduction into service in December 1927. Classmate No. 46167 THE HERTFORDSHIRE REGIMENT can also be seen ready for its days work with another unidentified 'Royal Scot' also in view. RCR5599

BELOW: 16TH FEBRUARY 1958 • 'Black Five' No. 45379 of Crewe North (5A) ambles past Camden shed with Fowler Class 3F 'Jinty' No. 47529 on the left. A few 3F tanks were allocated to 1B, including 47529 which remained there until November 1960. RCR5623

20TH SEPTEMBER 1958 ♦ 'Princess Royal' Class Pacific No. 46207 PRINCESS ARTHUR OF CONNAUGHT passes Camden shed with a Down express. RCR5557

THE R.C. RILEY COLOUR COLLECTION • MIDLAND REGION

3RD OCTOBER 1959 • Fowler Class 3P 2-6-2T No. 40053 storms up Camden Bank with an empty stock movement out of Euston. About a mile outside of Euston, the location pictured is at the top of a long 1 in 100 climb out of the West Coast Main Line's London terminus. **RCR5857**

THE R.C. RILEY COLOUR COLLECTION • MIDLAND REGION

31st March 1962 • Stanier 'Jubilee' No. 45660 ROOKE on the turntable at Camden shed. The loco is being turned on the 70 foot vacuum operated turntable installed in 1936 during the modernisation of Camden shed. The allocation under both the LMS and BR consisted almost entirely of Patriots, Jubilees, Royal Scots, Princess Royals and Princess Coronation Pacifics. In the mid-1950s no shed had a greater number of Coronation Pacifics than Camden. RCR5651

3RD JUNE 1962 ♦ Rebuilt 'Royal Scot' No. 46157 THE ROYAL ARTILLERYMAN passing Primrose Hill station on a parcels working. **RCR5611**

THE R.C. RILEY COLOUR COLLECTION • MIDLAND REGION

28TH MAY 1959 • Introduced in 1885, Dugald Drummond Caledonian Class 0F 0-4-0ST No. 56025 is pictured in St. Rollox Works yard, the place where it spent the majority of its working life until its withdrawal in 1960. No. 56025 was an early '264' class built in 1890. Later examples were built at St. Rollox railway works under the direction of John F. McIntosh from 1895 to 1908, these being given a different class number – 611. These small shunting locos had long service lives under the LMS and BR, with the last of the class withdrawn in 1962. Referred to by the generic term 'Pugs', they were mainly used as works shunters in the area around Glasgow, often running with home-made tenders to improve their small coal capacity. Like most 0-4-0 tanks of the period they had outside cylinders and inside slide valves driven by Stephenson valve gear. R.C. RILEY COLLECTION • RCR6147

16TH JUNE 1962 • Having not been in service for a while, as evidenced by the rust on the buffers and wheels, No. 54466 awaits its inevitable fate at Inverness shed. Built at St. Rollox Works for the Caledonian Railway in May 1916, this Pickersgill 4-4-0 gave over 45 years service being withdrawn in March 1962, eventually being scrapped in November 1963.
R.C. RILEY COLLECTION RCR5726

14TH SEPTEMBER 1959 ♦ Dugald Drummond Caledonian Railway 4-2-2 No. 123 and Highland Railway Jones Goods 4-6-0 No. 103 double-head a Scottish Industries Exhibition Special away from Renfrew Fulbar Street. A series of specials were run in conjunction with The Scottish Industries Exhibition held at Kelvin Hall in Glasgow between 3rd and 19th September 1959.

The service shown departed from Renfrew Wharf at 14.00 calling at Renfrew Fulbar Street, Paisley Abercorn, then ran via Rutherglen Triangle to Glasgow Central Low Level and Kelvin Hall. J.G. DEWING, R.C. RILEY COLLECTION ♦ RCR1709

ABOVE: 27TH MAY 1959 • Hughes Class 5MT 2-6-0 'Crab' No. 42743 at Kilmarnock station with a local passenger service. The 1927 Crewe-built loco was based locally at Hurlford shed (67B) from January 1952 until withdrawal at the end of December 1962. Kilmarnock station was opened on 20 July 1846 by the Glasgow, Paisley, Kilmarnock and Ayr Railway; this was connected to Ardrossan via Irvine two years later and to Carlisle via Dumfries & Gretna Junction in 1850. The current route to Glasgow (via Barrhead) was completed in 1871 jointly by the G&SWR and Caledonian Railway. RCR5690

RIGHT: 28TH MAY 1959 • Princess Coronation Class 8P 4-6-2 No. 46223 PRINCESS ALICE makes for a fine sight while at the head of the Up 'Royal Scot' at Glasgow Central station. Opened by the Caledonian Railway on 1st August 1879, Glasgow Central station is the northern terminus of the West Coast Main Line. Situated on the north bank of the River Clyde, it had eight platforms and was linked to Bridge Street station by a railway bridge over Argyle Street and a four-track railway bridge, built by Sir William Arrol, which crossed the Clyde to the south. However, the station soon became congested. In 1890, a temporary solution was found by widening the bridge over Argyle Street and inserting a ninth platform. Between 1901 and 1905 the original station was rebuilt by being extended over the top of Argyle Street, thirteen platforms were built. An additional eight-track bridge, the Caledonian Railway Bridge, was built over the Clyde, and the original bridge was raised by 30 inches. During the rebuild, a series of sidings was created at the end of Platforms 11 and 12 on the bridge over the River Clyde. These were named West Bank Siding, Mid Bank Siding and East Bank Siding. RCR5559

LEFT: 12TH AUGUST 1961 • Black Five 4-6-0 No. 44898 pictured at its lifetime home of Carlisle Kingmoor. Visiting the shed is Polmadie (66A) allocated 'Coronation' No. 46231 DUCHESS OF ATHOLL with a full tender of coal, presumably ready to return home to Glasgow. RCR5632

ABOVE: 16TH JUNE 1963 • 'Coronation' Class 8P Pacific No. 46242 CITY OF GLASGOW at Kingmoor shed. Mainly based at Camden during a working life of 23 years and 5 months, apart from short spells at Crewe, her final allocation was at Polmadie (66A) from where withdrawal took place on 18th October 1963 after clocking up more than 1.5 million miles. R.C. RILEY COLLECTION • RCR5582

The original Carlisle Kingmoor shed was built in 1876 and was of timber construction. During World War One there was a vast increase in traffic leading to the Caledonian Railway making the decision to rebuild the facility. Completed in 1916 with increased capacity and built of brick, it was now an eight-road through shed with a repair section. Additional facilities included a detached dormitory, canteen and wash house to accommodate those railwaymen who had completed a shift and were far from home.

The shed was located on the east side of the main line 1½ miles from Citadel station. Coded by the LMS as 12A until June 1950, it then transferred to the Scottish Region of British Railways and became 68A. February 1958 saw it return to the Midland Region and its previous code of 12A, closing in December 1967.

The current depot is located on the opposite side of the West Coast Main Line to the original steam shed and was officially opened on 1st January 1968.

10TH AUGUST 1963 • Ivatt-designed Class 2MT 2-6-0 No. 46455 is pictured on station pilot duties at Carlisle Citadel station. Allocated to Carlisle Upperby until 31st December 1966 it would end its days at Kingmoor before being withdrawn in May 1967.

In September 1847, the first services departed the Carlisle Citadel station, even though construction was not completed until the following year. The name derives from its location adjacent to Carlisle Citadel, a former medieval fortress. The station was built in a neo-Tudor style to the designs of English architect William Tite and is Grade II listed. Carlisle station was one of a number in the city; the others were Crown Street and London Road, but it became the dominant station by 1851. The other stations had their passenger services redirected to it and were closed. Between 1875 and 1876, the station was expanded to accommodate the lines of the Midland Railway which became the seventh railway company to use it.

The station is located on the West Coast Main Line, 102 miles south-east of Glasgow Central and 299 miles from London Euston. It is the northern terminus of the Settle and Carlisle Line, a continuation of the Midland Main Line from Leeds, Sheffield and London St. Pancras. It was formerly the southern terminus of the partially-closed Waverley Route from Edinburgh. RCR5706

21st August 1980 • Not a train in sight but the arches of Arten Gill viaduct, carrying the line over Artengill Beck, look impressive in this view. Located near Dent it comprises eleven arches, each of which are 45 feet across, it is 117 foot from the base at its highest point and is 220 yards long.

While largely constructed of sandstone, the inner arches of the viaduct spans were made using large blocks of Dent 'marble' (a type of dark limestone), from the now-disused quarries nearby, this stone was popular for use in ornamental masonry. The viaduct was originally designed to be further west, which is lower down the steep valley side, but by moving the line slightly eastwards, the viaduct could be installed at a higher location, thereby using fewer materials in its height. Groundwork on site was started in May 1870, with work on the viaduct itself beginning a year later, on 3rd May 1871.

Due to the loose soil and rock on the valley floor, some of the viaduct piers are rooted to a depth of 55 feet. The viaduct was constructed by Benton and Woodiwiss as part of Contract No. 2 (Dent Head to Smardale Viaduct, a total of 17 miles), Arten Gill is listed as bridge number 84 and, as with many of the viaducts along the line, was designed by John Sydney Crossley.

As indicated by the name of the viaduct, it crosses a steep valley through which the small watercourse of Artengill passes, but also an old pack-horse route eastwards across the moors. The route between Blea Moor Tunnel and Garsdale traverses the western edge of Great Knoutberry Hill and so needed several cuttings. These were prone to collecting snow during heavy winters, and attempts in clearing the snowdrifts involved dispatching the snow over the edge of Arten Gill viaduct.
RCR1755

4TH MARCH 1961 • Stanier 'Jubilee' No. 45569 TASMANIA entering the eastern end of Birmingham New Street station with a Down express.
RCR5646

4TH MARCH 1961 • Another Down express arrives at Birmingham New Street, this time with a rather unkempt rebuilt 'Patriot' at its head, No. 45521 RHYL.
RCR5670

4TH MARCH 1961 • Looking impressive in its clean BR black lined livery, Saltley-based Fowler Class 4MT 2-6-4T No. 42417 is seen in the early afternoon sunshine departing Birmingham New Street station with the 12.15pm local train to Worcester Shrub Hill. The loco is one of the final batch built which are identifiable by their side window cabs as modified from the original Fowler design by W. A. Stanier. RCR5855

New Street station was built in central Birmingham by the London and North Western Railway between 1846 and 1854, on the site of several streets in a marshy area known as 'The Froggery'; it replaced several earlier rail termini on the outskirts of the centre, most notably Curzon Street, which had opened in 1838 and was no longer adequate for the level of traffic. Samuel Carter, solicitor to both the LNWR and the Midland Railway, managed the conveyancing.

In 1846, the LNWR had obtained an act of Parliament to extend their line into the centre of Birmingham, which involved the acquisition of 3 acres of land and the demolition of around 70 houses in Peck Lane, The Froggery, Queen Street and Colmore Street. The Countess of Huntingdon's Connexion chapel, on the corner of Peck Lane and Dudley Street, which had only been built six years before, was also demolished. The station was formally opened on 1st June 1854, although the uncompleted station had already been in use for two years as a terminus for trains from the Stour Valley Line that entered the station from the Wolverhampton direction. On the formal opening day, the LNWR's Curzon Street station was closed to regular passenger services and trains from the London direction started using New Street.

The LNWR originally shared the station with the Midland Railway; however, in 1885, the Midland Railway opened its own extension alongside the original station for the exclusive use of its trains, effectively creating two stations side by side. The two companies' stations were separated by a central roadway, Queens Drive. Traffic grew steadily and, by 1900, New Street had an average of 40 trains an hour departing and arriving, rising to 53 trains in the peak hours.

Midland Railway trains that had used Curzon Street began to use New Street from 1854; however, its use by the Midland Railway was limited by the fact that those trains going between Derby and Bristol would have to reverse, many trains bypassing New Street and running through Camp Hill. This was remedied in 1885, when a new link to the south, the Birmingham West Suburban Railway, was extended into New Street, which allowed through trains to and from the south-west to run via New Street without reversing.

THE R.C. RILEY COLOUR COLLECTION • MIDLAND REGION

THE R.C. RILEY COLOUR COLLECTION • MIDLAND REGION

16TH APRIL 1955 • A busy scene at Bromsgrove South as Fowler 4F 0-6-0 No. 44133 waits outside the signal box waiting for permission to proceed.

On the left there's an unidentified Fowler 'Jinty' at the coaling stage while further along the banking loco siding sits 'Big Bertha'. In the distance close to the gantry just outside Bromsgrove station there seems to be plenty of activity taking place.

The railway reached Bromsgrove in June 1840 when the first section of the Birmingham & Gloucester Railway from Cheltenham opened. From Bromsgrove passengers had to continue their journey north by coach until in September of the same year the next section of the railway was opened to Barnt Green; this brought into play the two miles of the 1 in 37.7 Lickey Incline. The line reached Birmingham on 17th December 1840.

Bromsgrove station was built in a cutting on the east side of the town, with three lines running through the station. The centre road was initially used for banking engines but later was used as the Down fast road. RCR1763

16TH APRIL 1955 • Midland-built 'Lickey Banker' 0-10-0 No. 58100 at Bromsgrove South gets ready for its next duty. 'Big Bertha' (its Derby nickname), as she was commonly known, was built in 1919 by the Midland Railway. Originally numbered 2290 (changed later by the LMS to 22290 in 1947 to avoid duplication with a new build Fairburn 2-6-4T). It was designed by James Anderson specifically for banking duties on the Lickey Incline, the job it did until withdrawal in May 1956, reputedly due to a replacement boiler and/or firebox being required, both of which were of course non-standard and consequently it would have been hard to justify the cost. No. 58100 was the largest locomotive the Midland Railway ever constructed and was scrapped where she was built, at Derby Works, in September 1957.

There is a degree of uncertainty why she was called 'Big Bertha'. It certainly seems a strange copy of the name given to a huge German gun based in Belgium in WWI which was able to reach the English shores. This is especially so with the origins of the name known to many railwaymen. Whatever, it was said with a degree of affection but was only ever an unofficial title, no name was ever carried. Perhaps because of the continental connection she was also referred to as 'Big Emma'. RCR5727

20TH APRIL 1957 ♦ Waiting for their next duties, Johnson Class 3F 0-6-0 No. 43762 is pictured alongside classmate No. 43186 and 'Jinty' No. 47502 at Bromsgrove Shed. Banking engines were required at Bromsgrove from the outset when the Lickey Incline opened in 1840. The shed was built by the Birmingham and Gloucester Railway alongside their works to service the Lickey banking engines, with additional buildings added over the years. The shed remained in use until September 1964 when the banking engines were made redundant by the introduction of diesels on the Birmingham to Bristol main line. **RCR5776**

20TH APRIL 1957 • Fowler Class 3F 0-6-0T 'Jinty' No. 47638 makes its way towards the rear of a waiting train at Bromsgrove station that will soon be heading up the Lickey Incline. As can be seen in this photograph the signal box was located on the Up platform and was constructed at a lower height so visibility through the road bridge at the end of the platform was not compromised too much. Although not obvious from this angle it is set forward from the building line of the station offices to provide a clear view along the platform. The station was built on an ascending gradient of 1 in 300 so the engines would already have been working hard before hitting the gruelling section of the climb.

In LMS days no train was permitted to tackle the Lickey Incline without a banking engine other than freight trains of not more than eight wagons and passenger trains not exceeding more than six carriages. RCR5906

24TH APRIL 1957 • Fowler Class 4F 0-6-0 No. 43971 shunting at Stratford-upon-Avon Old Town station. Note that the signal box board has a slightly different spelling with the middle word losing its 'up' to become Stratford-on-Avon.

The Stratford-upon-Avon & Midland Junction Railway (SMJR) was one of Britain's more impoverished and least efficient small railways. Its line ran across a largely empty, untouched portion of England visiting the counties of Northamptonshire, Warwickshire, Oxfordshire and a small part of Buckinghamshire. It was noted that 'within the county of Warwickshire the railway had a route mileage of 28 miles'. It only existed as the SMJR from 1909 to 1923 when it became part of the LMS. Its origins lie in the amalgamation of four railway companies in 1909-10: the Northampton & Banbury Junction Railway (NBJR), the East & West Junction Railway (E&WJR), the Evesham, Redditch & Stratford-upon-Avon Junction Railway and the Easton Neston Mineral & Roade and Olney Junction Railway. In 1919 its route mileage was 67 miles 46 chains. RCR1886

29TH APRIL 1956 • Johnson Class 3F 0-6-0 No. 43222 waits for its passengers to return at Stratford-upon-Avon Old Town station during a Stephenson Locomotive Society rail tour. There are a few interested observers, including the signalman, to watch the train depart.

The sparse passenger traffic on this rural route meant that the line struggled to be profitable and on 5th April 1952 regular passenger services between Blisworth and Stratford-upon-Avon Old Town ceased. Freight traffic would continue until June 1965, locos for this purpose can be seen in the shed on the right of the picture.

On departure the tour would head for Birmingham New Street where the veteran 3F, dating from 1890, would hand over to Standard 5MT No. 73099 for the return to the capital. RCR5790

S.M.J.R. RAIL TOUR – M500 – 29th April 1956 Stephenson Locomotive Society	
62605 Class D16/3	London Kings Cross – Hertford North – Hitchin
43222 Johnson Class 3F	Hitchin – Bedford – Olney – Stoke Bruern – Towcester – Moreton Pinkney – Byfield – Kineton – Stratford (SMJ)
43222	Stratford (SMJ) – Binton – Broom Junction East – Broom Junction North – Alcester – Coughton – Redditch – Barnt Green – Selly Oak – Birmingham New Street
73099 Standard 5MT	Birmingham New Street – Canley – Coventry – Kenilworth – Leamington Spa – Marton Junction – Dunchurch – Rugby Midland – Weedon – Bletchley – Watford Junction – London Euston
Notes:	It had been hoped to have a Clan Pacific for the final leg but none were available.

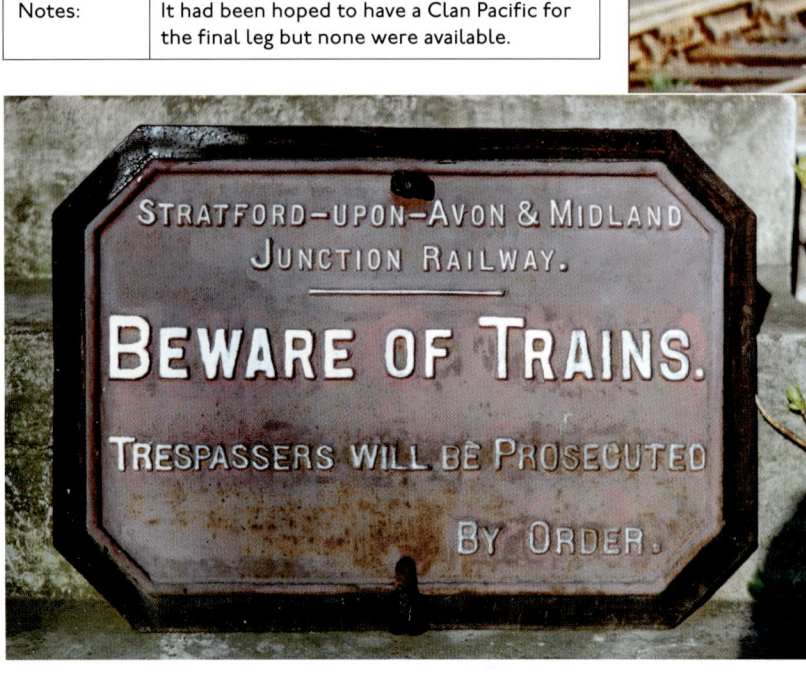

16TH FEBRUARY 1958 •
SMJR Beware of Trains notice. RCR2126

24th April 1957 • The driver of Fowler 4F 0-6-0 No. 43971 looks towards the camera after dropping off the rear of a departing freight from Stratford-upon-Avon Old Town. The freight is heading towards Broom Junction and is being hauled by fellow 4F No. 43876. The delightful and rural nature of the Stratford-upon-Avon & Midland Junction Railway is clear to see in this superb image. RCR5788

THE R.C. RILEY COLOUR COLLECTION • MIDLAND REGION

THE R.C. RILEY COLOUR COLLECTION • MIDLAND REGION

ABOVE: 27TH APRIL 1963 • Stanier Class 8F No. 48751 pictured working an Up freight near Banbury. RCR5766

LEFT: 5TH DECEMBER 1959 • External and internal views of Banbury Merton Street station. RCR1762/RCR1761

Banbury Merton Street was the first railway station to serve the Oxfordshire market town. It opened in 1850 as the northern terminus of the Buckinghamshire Railway, providing connections to Bletchley and Oxford and closing for passengers in 1961 and goods in 1966. The station was the northern terminus of the Buckinghamshire Railway which consisted of two lines: one from Bletchley to Banbury and another from Verney Junction to Oxford. Construction of the line had begun in July 1847 but was beset by delays and financial problems; priority was given to the construction of the line to Banbury and this was completed on 30th March 1849, with the section from Claydon to Banbury being built to single track rather than double, as had been intended. The Oxford branch was opened on 1st October 1850 as far as Islip, reaching a temporary station at Oxford Road on 2nd December.

The line was worked from the outset by the London and North Western Railway (LNWR) which had supported the building of the line, eventually being absorbed by the LNWR on 23rd February 1878. Merton Street reached its peak of passenger and goods traffic at the outbreak of the First World War, this included handling troop trains converging from north to south. In 1916, the Ministry of Munitions constructed a National Filling Factory on the northern side of the line. This closed in 1919 but was then converted into a factory to break down thousands of tons of war materials, a process which continued until 1924.

Merton Street saw a growth in freight traffic, but unfortunately this was matched by a fall in passenger numbers. By 1938, the LMS planned to phase out Merton Street by agreeing with the Great Western to rebuild the two stations as a single unit situated to the north of the present Banbury station. Owing to the outbreak of the Second World War, this plan was never put into action and Merton Street was once again busy with troop trains.

The post-war period saw a further decline in passenger numbers, however freight receipts remained steady as a result of cattle traffic. Around 200 cattle vans were handled per week, these activities continued until the early 1960s when British Railways began to phase them out.

By 1956 the timber boarding on the station roof had reached such a condition that it posed a danger to passengers and it was removed leaving the metal supports and piping which, as seen in our images, were painted white. In August of the same year steam-hauled services were replaced by DMUs, unfortunately this had not succeeded in stemming the line's losses and a proposal to withdraw passenger services was published in July 1960, with the last passenger train running on 31st December.

13TH MAY 1965 • Stanier 8F 2-8-0 No. 48404 on a Down freight near Claydon on the former 'Varsity Line' from Oxford to Cambridge. The station was situated between Steeple Claydon and Middle Claydon in Buckinghamshire.

Built at Swindon Works and into traffic on 23rd July 1943, in the British Railways era No. 48404 was one of a number of 8Fs allocated to the Western Region of which 22 were modified as they passed through the works for general repairs to have GWR-pattern ATC equipment fitted, as can be seen at the front of the boiler barrel. The work was mainly carried out at Horwich Works. 48404's was fitted in October 1957. RCR5768

25TH AUGUST 1962 • Stanier Mogul No. 42970 with a freshly filled tender is pictured at Oxford Rewley Road shed.

Construction of the first Rewley Road shed was approved in February 1851; when built it was predominantly a corrugated iron construction comprising three roads. In October 1877 disaster struck when the shed was blown down. It was patched up but after a few years it was beyond a state of repair and a new two-road brick built shed was approved and constructed, opening in late 1882. In 1910 its original 42 foot turntable was replaced by a 50 foot version and engines were coaled for almost all of its existence from open wagons. Although covered shelter was provided in March 1950 this was only a short-lived facility as in December of that year engines stopped using the shed and were serviced at the ex-GWR shed opposite. From LMS days until its closure it was a sub-shed of Bletchley (2B). RCR5692

ABOVE: 14TH SEPTEMBER 1963 • Fowler Class 4F 0-6-0 No. 44092 takes a breather in the loop north of Ashchurch station with a lengthy mixed freight working. The first two wagons, four-wheel flatbeds, are transporting BMC Morris cab and chassis while the rest of the consist appears to be empty mineral wagons with a few box vans towards the rear. The footplate crew look out for the train they have been looped for. RCR5791

RIGHT: 18TH JULY 1965 • A busy scene outside Ashchurch signal box as Fowler 4F 0-6-0 No. 44560 stands on the Down line, while a trio of locomotives, Standard Class 5MT 4-6-0 No. 73053, Stanier Class 8F 2-8-0 No. 48120 and a second Standard Class 5MT, No. 73032, make their way through Ashchurch station.

The locos have probably been involved with MoD workings; the line into the depot runs behind the signal box and the small brick building to the left of the 4F.
R.C. RILEY COLLECTION • RCR5817

BELOW: 19TH JULY 1958 ♦ At the quiet little Worcestershire village station of Ripple, a popular destination for fishermen, Stanier Class 2P 0-4-4T No. 41900 pauses with an Upton-on-Severn to Tewkesbury service.

Ripple station opened under the Midland Railway on a newly constructed 14 mile branch line linking their Birmingham to Ashchurch line with the GWR's Worcester to Hereford line in 1861. The line headed westwards to Tewkesbury, then bridged the River Severn to Upton-on-Severn and Malvern Wells, where it made its connection with the GWR line. At the time the photograph was taken the station only saw three weekday and five Saturday trains. Along with the branch itself the station led a fairly inauspicious life in the region's railway history, being reduced to single-line working following the truncating of the route with the closure of the Malvern to Upton-on-Severn section in 1952. The station, along with the line, finally closed to passengers on 12th August 1961. RCR5891

ABOVE: 22ND JULY 1961 • Ex-Midland Railway Johnson Class 3F 0-6-0 No. 43754 pictured at Ashchurch waiting to depart with an Upton-on-Severn service.

The building of the original Ashchurch station was authorised in 1836 to the Birmingham and Gloucester Railway, whose central section from Bromsgrove to Cheltenham, which included Ashchurch, was opened on 24th June 1840; the rest of the line was opened throughout a few months later and it subsequently became part of the Midland Railway. Ashchurch was a beautifully constructed junction station and was the junction for two branches, one each side of the main line: the Evesham loop line and the line to Tewkesbury, Upton-on-Severn and Great Malvern.

The station was closed in November 1971 with almost all of the buildings being demolished – only one small derelict red brick shed remained of the original station buildings. However, on 1st June 1997 the station reopened on the site of the earlier one which had lain derelict for 26 years. In the post-war period, the station had been used both for passenger services and for goods loading for the nearby army base; a number of sidings still exist nearby. RCR5781

ABOVE: 19TH OCTOBER 1959 • Wolverhampton High Level station is the location for departing Stanier 'Jubilee' No. 45734 METEOR. When the station opened on 24th June 1852 it was known as Wolverhampton General but was quickly renamed Queen Street station in September 1853; the entrance faced onto Queen Street, so the name made it easier to locate. Work was in progress on the GWR station which was to become known as the Low Level, so Queen Street unsurprisingly became known as the High Level and was renamed so on 1st June 1885. The Midland Railway had used the station from 1st September 1867. Remaining largely unchanged through LMS and BR ownership, the station was demolished in January 1965 and rebuilt as part of the electrification scheme. RCR5659

OPPOSITE TOP: 28TH MAY 1960 • A view of the ex-LMR Wednesbury Town station on the South Staffordshire Line with the signalman seemingly having a relaxing time in the sunshine. RCR1914

OPPOSITE BOTTOM: 28TH MAY 1960 • No. 49275 shunting at Wednesbury. RCR5739

Wednesbury station was built by the London and North Western Railway in 1850. The station stood on the line linking Dudley with Walsall and at one time the junction with the branch serving the Darlaston loop. In 1923 the station came under the control of the LMS. As with numerous areas within the region, until 1950 there were effectively two Wednesbury stations with the addition of 'Town' being made to the London, Midland Scottish station and 'Central' appended to its Great Western Railway counterpart a few hundred yards away. Wednesbury Town was closed in 1964 leaving the town to be served by its counterpart located just a few hundred yards away.

THE R.C. RILEY COLOUR COLLECTION • MIDLAND REGION

23RD APRIL 1960 • Hughes Class 5MT 2-6-0 No. 42941 on shunting duties at Wellington, the coal piled high in the Fowler-designed tender doesn't look the best quality does it? RCR5689

In 1845 Wellington was a quiet market town but as 'Railway Mania' dawned, the next 20 years would see the railway transform the area. This was mainly due to the rivalry of the GWR and LNWR and by 1870 lines would radiate in five different directions from the town. Both companies would build warehouse facilities and provide a network of sidings to service local industries with connections to Coalbrookdale, the cradle of the industrial revolution.

RIGHT: 27TH JUNE 1964 • Black Five No. 44852 on a Newton Abbot–Bradford train departing Gloucester Eastgate station past the Midland Type 3b signal box which opened with the line on 12th April 1896, closing on 24th March 1968.

Gloucester Eastgate station was originally a terminus station on the Birmingham and Gloucester Railway. In 1896 a new through station was constructed on the Tuffley Loop line on a site that had previously housed the Midland Railway engine shed. The station, originally known simply as Gloucester, opened on 12th April 1896; it had three through platforms and one bay. It was sharply curved and featured a main entrance building as well as buildings on both platforms. The buildings were of typical Midland Railway design in red brick with terracotta decoration. Extensive cast-iron and glass ridge-and-furrow awnings covered the platforms. There was a direct footbridge connection to the GWR's Gloucester Central.

The station was renamed Gloucester Eastgate by British Railways on 17th September 1951. After 1968, the station was rationalised. The island platform was lengthened at the southern end and the tracks were removed from the other two platforms. The extensive goods yard and sidings were also lifted at this time. Colour-light track-circuit block signalling was installed, and the station was effectively merged with Gloucester Central.

Eastgate station was closed on 1st December 1975 along with the Tuffley Loop. This was partly an attempt by British Rail to cut maintenance costs and partly a result of pressure from the road lobby and local councillors who wanted to rid Gloucester of four of its level crossings. Services that had previously called at Eastgate now had to perform a reversal at Central station, an operational inconvenience that led to fewer trains calling at Gloucester. The administrative offices on Eastgate station lingered on in use for nearly two more years until demolition came in 1977. R.C. RILEY COLLECTION • RCR5616

4TH JUNE 1962 ♦ Fowler Class 4F 0-6-0 No. 44272 heads north near California Crossing, Gloucester on the Tuffley Loop. The train is comprised of steel mineral empties with a couple of older wooden bodied wagons near the head of the train. No. 44272 was allocated to Gloucester Barnwood (85C) at this time but would shortly move to Templecombe (82G) on the S&D in August and was taken into storage there in April 1963. **RCR5819**

THE R.C. RILEY COLOUR COLLECTION • MIDLAND REGION

ABOVE: 4TH JUNE 1962 • Fowler Class 4F 0-6-0 No. 43924 makes its way through Gloucester Park with a short freight from High Orchard Sidings. This was another Barnwood allocated locomotive, built in 1920 by the Midland Railway, therefore a right-hand drive version. The tall chimney and factory buildings in the background were part of Fielding & Platt's extensive engineering works; it no longer stands and the route of the railway now forms part of the A430 Trier Way. RCR5777

BELOW: 4TH JUNE 1962 • We are a little closer to the engineering works to see Deeley 0-4-0T No. 41535 pushing a train made up of mainly wooden bodied wagons back towards High Orchard Sidings. RCR6070

4TH JUNE 1962 • No. 41535 has paused on the approach to Llanthony Road crossing after picking up its short train of vans from Victoria Dock where the driver will wait for the fireman to signal it is safe to proceed over the crossing. The line curving off to the left led to the ex-GWR side of the docks over a swing bridge. RCR6066

4th June 1962 • Deeley Class 1528 0-4-0T No. 41535 pictured on shunting duty in Merchants Road, Gloucester. This shunting locomotive was built by the Midland Railway in 1897 at Derby Works and numbered 1137A, being renumbered 1521 in 1907 and then 1535 in February 1922. Ten of these small shunting locomotives were built in two batches of five, the second batch with only minor detail differences. Withdrawal of the class was gradual beginning in 1957, the final two (41528/41533) being withdrawn in 1966.

No. 41535 was based at Gloucester Barnwood (85C) until the end of September 1963, being reallocated to Swansea East Dock (87D) for a short spell before ending its working life of 67 years and 9 months at Neath (Court Sart, 87A), being withdrawn from there on 14th September 1964.

The train is standing outside the premises of Associated Electrical Industries (A.E.I.), a large national lamp and lighting company that was later acquired by the General Electric Co. which would then create the UK's largest electrical group at the time. Merchants Road and the surrounding industrial area were served by an extensive and complicated network of lines, all established when the railways had little competition. The siding pictured originally ran the whole length of Merchants Road but was gradually shortened from the early 1960s. The area has undergone redevelopment but the buildings pictured are still in use today. RCR6062

Mangotsfield station was situated approximately five miles north of Bristol Temple Meads. The original station was opened in 1845 by the Bristol and Gloucester Railway, but was resited in 1869 to serve the new Mangotsfield and Bath branch line, and became an important junction station with extensive facilities and a total of six platforms. Passenger usage never matched the size of the station; at its peak eight staff were employed.

The 'new' station was built by the Midland Railway at the time of the opening of the Bath to Bristol line on 4th August 1869, replacing the original Bristol and Gloucestershire Railway station at Mangotsfield North Junction. It was constructed on the southern face of Rodway Hill (visible on the left of main picture), having three island platforms, thus giving six platform faces with the junction apex forming the western end of the middle island as seen in the picture below.

The building to the left of the locomotive is Carson's Chocolate factory which was opened in the middle of the Mangotsfield railway triangle in the early 1910s. Served by a private siding, and with its own cricket pitch, tennis courts and bowling green, the factory became a well-known landmark for rail travellers, as well as an extra source of passengers for the station.

The station closed on 7th March 1966 alongside the withdrawal of Bath Green Park to Bristol passenger services coinciding with the closure of the S&D.

5TH OCTOBER 1962 • Two views of Mangotsfield station with rebuilt 'Royal Scot' No. 46117 WELSH GUARDSMAN departing with a parcels train heading in the Bristol direction. RCR5606/RCR1996

ABOVE: MARCH 1965 • A very clean Stanier Class 8F 2-8-0, No. 48309, at Bath Green Park shed. The reason for the state of the locomotive is that it was involved in a leg of an LCGB rail tour 'The Wessex Downsman (No. 1)' on 4th April 1965. 48309 would take the tour from Bath Green Park via the S&DJR to Bournemouth West.

Both the Midland Railway and the Somerset and Dorset Joint Railway opened locomotive depots near the station on the west side of the River Avon. The Midland Depot opened in 1869 and the S&DJR in 1874, both depots closing in March 1966. RCR5743

ABOVE RIGHT: 28TH APRIL 1957 • Ivatt 2MT 2-6-2 tanks Nos. 41202 and 41203 are pictured at Yatton station in north Somerset on the GW Bristol to Exeter line. They are working the RCTS (London Branch) North Somerset Rail Tour.

The tour started at London Waterloo, working via Reading to Bristol Temple Meads where the Ivatts took over to work the train to Highbridge. 41202 then took the train on a return trip to Burnham-on-Sea before 41203 rejoined the train again on its journey back to Temple Meads; here they left the train which was hauled back to London Paddington by No. 3440 CITY OF TRURO and Class 45XX 2-6-2T No. 5528. RCR5870

BOTTOM RIGHT: 25TH AUGUST 1962 • Stanier 'Black Five' 4-6-0 No. 44817 at Bristol Temple Meads on a southbound parcels train. A wartime engine built at Derby in 1944, it was shedded at Burton from April 1962 until May 1963. RCR5617

THE R.C. RILEY COLOUR COLLECTION • MIDLAND REGION

THE R.C. RILEY COLOUR COLLECTION • MIDLAND REGION

ABOVE LEFT: 5TH OCTOBER 1962 • Bristol Barrow Road shed is the location for this shot of Fowler 4F 0-6-0 No. 44092 and ex-L&Y 'Pug' No. 51218, which is being drawn out of the yard towards Barrow Road Bridge (which bisected the yard) by the 4F. At the time the photo was taken Barrow Road had two of the ex-L&Y 'Pugs' allocated to it which were used for shunting on the restricted Avonside Wharf branch. Under the LMS the shed was coded 22A but changed to 82E when placed under Western Region jurisdiction by BR. The shed closed in October 1965.

Built by Kerr, Stuart to an LMS order, No. 44092 dates from November 1925, being withdrawn by BR in September 1964. No. 51218 was built in 1901 at the Lancashire and Yorkshire Railway's Horwich Works as No. 68. Acquired direct from British Railways in late 1964 and delivered directly from Neath, South Wales, where it had ended its working days in June of the same year, 51218 went to the Keighley and Worth Valley Railway, becoming the first locomotive to arrive there on 7th January 1965. RCR5807

BELOW LEFT: 5TH JULY 1959 • An earlier view of Barrow Road shed with Fowler 'Patriot' 4-6-0 No. 45506 THE ROYAL PIONEER CORPS in the company of an unidentified Churchward Mogul. This was one of three 'Patriots' allocated to 82E, the others being Nos. 45504 and 45519, that were used on cross-country services from the south-west to northern cities, including 'The Devonian'. RCR5671

ABOVE: 23RD OCTOBER 1965 • Fowler Class 3F 'Jinty' 0-6-0T No. 47276 pictured at Bristol Bath Road shed which was built originally by the Bristol and Exeter Railway and opened in January 1852.

Bath Road mainly handled passenger traffic locomotives, while St. Philip's Marsh depot on the eastern throat handled freight types. Post-nationalisation, under British Railways both Bath Road and St Philip's Marsh gained additional locomotives from the closure of the local Midland Region sheds; by 1950 it had an allocation totalling 93 locomotives.

Bath Road was one of the first sheds to be closed to steam locomotives from September 1960. Rebuilt as a diesel depot (note the Blue Pullman to the rear of the Jinty), it retained one of the turntables. The depot ceased all operations on 2th September 1995. RCR5903

14TH APRIL 1962 • Fowler Class 2P 4-4-0 No. 40646 double-heads with another Fowler-designed engine, Class 3P 2-6-2T No. 40026, arriving at Northampton Bridge Street with rail tour 1X47, 'Tour of Seven Branch Lines'. Note that 40026 is one of the class fitted with condensing gear for working on the Moorgate widened lines.

The station was originally named Northampton, being the first station serving the town. It opened in 1845, subsequently renamed Northampton Bridge Street in 1876 after a new station was built for the line to Market Harborough. The station meant that people could travel to Wellingborough, Irthlingborough and Peterborough more quickly than before.

The station closed to passengers in 1964, the buildings being demolished in 1969. Freight trains continued to use Bridge Street regularly until 1972; a lone remaining group of private sidings finally closed in 2005.

R.C. RILEY COLLECTION • RCR5720

TOUR OF SEVEN BRANCH LINES – 1X47 – 14th April 1962 Stephenson Locomotive Society (Midland Area)	
40646	Birmingham New Street – Whitacre Junction – Nuneaton Abbey Street – Nuneaton Trent Valley – Bedworth – Three Spires Junction – Humber Road Junction – Rugby – Long Buckby – Northampton Castle
40646 40026	Northampton Castle – Northampton Bridge Street – Hardingstone Junction – Ravenstone Wood Junction – Olney – Bedford
40026	Bedford – Shefford – Hitchin
1247	Hitchin – Stevenage – Hertford North – Cole Green – Welwyn Garden City – Hatfield
1247	Hatfield – Welwyn Garden City – Luton Bute Street
40646	Luton Bute Street – Dunstable North – Leighton Buzzard – Bletchley – Hanslope Junction – Weedon – Daventry – Marton Junction – Leamington Spa Avenue – Kenilworth – Berkswell – Stechford – Birmingham New Street

24TH OCTOBER 1959 • A view looking south of Coventry station, which is in the course of being rebuilt. A very busy service hauled by Stanier 'Black Five' 4-6-0 No. 45038 has recently arrived. RCR5686

The original Coventry station was built in 1838 as part of the London and Birmingham Railway. Within two years it had been replaced by a new larger station. The original station buildings remained in service as the station masters' offices until the station was redeveloped in the early 1960s by the London Midland Region of British Railways.

The first rebuilding of the station in 1840 saw a significant number of modifications and extensions over the years, additions included an engine shed, water column and turntable, and in its later days an inclined walkway from the platform directly to Warwick Road for summer excursions, plus a parcel depot formed from old carriages. However, the station was constrained by bridges at either end, Stoney Road Bridge to the south, and Warwick Road Bridge to the north. The bridges effectively restricted the station to two lines, and prevented the platforms from being extended. Starting in 1959, as seen in these pictures, the old station and both of the aforementioned bridges were demolished and rebuilt. The bridges were widened, and this made it possible to increase the station to four platforms to replace the previous two, shown in the image above. The rebuilt station was formally reopened on 1st May 1962.

THE R.C. RILEY COLOUR COLLECTION • MIDLAND REGION

LEFT: 28TH OCTOBER 1967 • Two views of Jubilee No. 45562 ALBERTA at Halifax on its leg of the Preservation Special. This was probably the last working of 45562 as she was out of use for three weeks prior to the tour, and withdrawn shortly after on 4th November from Holbeck (55A). RCR5643/RCR5644

ABOVE: 1960/1961 • Aspinall Class 1008 2-4-2T No. 50850 pictured at Southport on an unrecorded date, although as it was allocated to Southport shed (27C) between June 1960 until withdrawal in November 1961, the photo was probably taken between those dates.

Built at Horwich Works to order 38, No. 50850 was released to traffic as L&YR Class 5 No. 675 during September 1899. The primary task for these locomotives was for local passenger work. Across the 310 members of the class there were variations made, many were rebuilt with Belpaire boilers and four different cylinder sizes – $17\frac{1}{2}$, 18, $19\frac{1}{2}$ and $20\frac{1}{2}$ inches – were used. There were also superheated versions developed and these were rated as Class 6 locomotives. R.C. RILEY COLLECTION • RCR6148

PRESERVATION SPECIAL – 1Z75 – 28th October 1967 Manchester Rail Travel Society/Severn Valley Railway Society	
Electric	Birmingham New Street – Wolverhampton – Stafford – Crewe – Stockport
45411	Stockport – Heaton Norris Junction – Denton – Miles Platting – Manchester Victoria
70013	Manchester Victoria – Castleton – Rochdale – Hebden Bridge – Brighouse – Huddersfield – Penistone – Wadsley Bridge – Sheffield Victoria – Woodburn Junction – Rotherham Central – Swinton – Mexborough – Doncaster – Hare Park Junction – Turners Lane Junction – Normanton
45562	Normanton – Wakefield Kirkgate – Halifax – Low Moor – Laisterdyke – Ardsley – Wakefield Westgate – Normanton
70013	Normanton – Altofts Junction – Holbeck Low Level (Leeds) – Kirkstall – Keighley – Skipton – Carby – Colne – Rose Grove
73040	Rose Grove – Accrington – Blackburn – Darwen – Moses Gate – Salford Crescent – Manchester Victoria – Miles Platting – Denton – Heaton Norris Junction – Stockport
Electric	Stockport – (reverse of outward route) – Birmingham New Street

THE R.C. RILEY COLOUR COLLECTION • MIDLAND REGION

An unidentified Crab passes Chester No. 4 signal box and under the magnificent signal gantries on a local service approaching the western end of the LNWR/GWR joint station.

From 1875 to 1969 the station was known as Chester General to distinguish it from Chester Northgate. The station's Grade II* listed Italianate frontage was designed by the architect Francis Thompson and constructed by Thomas Brassey while engineer C. H. Wild designed the train shed.

Construction began on 1st August 1847, exactly a year later the station was officially opened to great acclaim by the city's populace due to the wide range of destinations that could be reached for the first time.

When built, the station had a single through platform, a pair of bay platforms, and the main building. Expansion of the station and its environs was required within the first few decades of opening, this comprised the construction of sidings, warehouses, signal boxes and motive power depots. To accommodate the increasing number of passengers and freight in the 1860s and 1870s, the station was extended again. Two island platforms, two bay platforms, and additional facilities were connected via a footbridge to the existing station, being completed by 1890. RCR2053

THE R.C. RILEY COLOUR COLLECTION • MIDLAND REGION

THE R.C. RILEY COLOUR COLLECTION • MIDLAND REGION

11TH JUNE 1961 • 6A Chester (Midland) shed was situated on the north side of the Chester-Crewe line, east of Chester General station and that is where we see Fowler 'Patriot' Class 6P 4-6-0 No. 45547 of Llandudno Junction (6G) shed pictured alongside rebuilt 'Royal Scot' No. 46158 THE LOYAL REGIMENT. Looking in good condition, a transfer to Edge Hill shed of the unnamed 'Patriot' was imminent, where the loco would see out its days until withdrawal in September 1962.
R.C. RILEY COLLECTION • RCR5649

TOP LEFT: 6TH AUGUST 1960 • One of the four named 'Black Fives', No. 45156 AYRSHIRE YEOMANRY, at Crewe station departing with a local stopping service. The engine was named on 19th September 1936 while allocated to the Northern Division of the LMS at Carlisle Kingmoor. Underneath the crest sits a separate straight plate that reads 'Earl of Carrick's Own'. The plates for all of the named Black Fives were cast and fitted at St. Rollox Works and replicated the same lettering style as produced on the Jubilees whose plates had also been made there. RCR5614

BOTTOM LEFT: 24TH SEPTEMBER 1961 • 'Princess Royal' Class 8P No. 46211 QUEEN MAUD pictured at Crewe South (5B), although looking in reasonable condition, she would be withdrawn from service two weeks after this photograph was taken.

Built at Crewe Works and entering service on 18th September 1935 from Camden shed (1B), 46211 amassed over 1.5 million miles during a service life of 26 years, withdrawal coming on 7th November 1961 and cutting up occurred at Crewe during April 1962. RCR5562

ABOVE: 25TH SEPTEMBER 1961 • No. 46170 BRITISH LEGION at Crewe. It was the prototype for the rebuilt Royal Scots, but differed from those later rebuilds principally in having a unique type 2 boiler, rather than a type 2A boiler, the two not being interchangeable. Although not strictly speaking a rebuilt Royal Scot, numerically it was the final member of the class.

In 1935 No. 6170 was constructed using the frames of the unsuccessful experimental high pressure compound locomotive No. 6399 FURY; it was also equipped with a William Stanier-designed new Type 2 taper boiler along with a new smokebox and inside cylinder. It remained the only Class 6P rebuild until 1942 when two Jubilee Class engines, Nos. 5735 COMET and 5736 PHOENIX, were rebuilt with 2A boilers. After that, the LMS rebuilt all 70 of the Royal Scots, between 1943 and 1955, and 18 of the Patriot Class between 1946 and 1949 with 2A boilers. When built, 6170 was given the standard passenger livery of LMS crimson lake. There were numerous differences between 6170 and the rebuilt Royal Scots, the boiler being longer, it had a single chimney, although as the photograph shows, a double chimney was fitted later; it was also equipped with a new cab and side windows. RCR5595

6TH AUGUST 1960 • A passenger standing in the adjacent DMU admires the elegant lines of 'Princess Royal' Class 8P No. 46205 PRINCESS VICTORIA at Crewe station. Built at Crewe Works, it entered service on 24th July 1935, allocated to Camden (1B). In 1938 No. 6205 underwent modification to the motion with the outside gear driving the inside valves via rocking levers, this replacing four sets of motion previously used. As can be seen in this photograph a large tripod shaped motion bracket had to be provided to handle the increased load on the outside valve gear.

6205 retained its LMS Crimson Lake livery until 1946 when it was repainted into LMS lined black and then became the first of the class to receive British Railways lined black in November 1948 (as 46205), retaining this until 1952 when the final livery of green was applied. During its lifetime the loco underwent a total of 17 heavy repairs, slightly less than the class average of 19½, and clocked up over 1.4 million miles, being allocated mainly at Edge Hill and Crewe North between 1947 and 1961, eventually being withdrawn from Willesden (1B) on 25th November 1961.

Post-war the 'Princess Royal' class were allocated to the Euston-Liverpool named expresses – 'The Merseyside Express', 'The Shamrock', 'The Red Rose' and the summer only service, 'The Manxman'. They were also regular visitors north of the border including regular turns on the West Coast Postal, hence the high mileages attained by all members of the class.
RCR5547

1ST SEPTEMBER 1964 • Stanier Class 5MT 4-6-0 No. 45298 at Crewe station. Built by Armstrong Whitworth in December 1937 and initially allocated to Edge Hill shed (8A). It is seen here attached to one of the two 1946-built coal-weighing tenders – No. 10591, originally paired with No. 4996. Built at Horwich, coal-weighing tenders were produced at the request of the Motive Power Department because of the rising cost and deterioration in the quality of coal. Their main purpose was to study the effects on coal consumption due to traffic delays and other operational features and also to check the effects of different sources of coal supply on a day-to-day working by monitoring different methods of firing. 45298 was paired with the tender on 30th August 1958 until withdrawal in September 1967, scrapping took place during February 1968 at Cohens (Kettering). RCR5679

5TH JUNE 1960 ♦ Rebuilt Class 7P No. 46100 ROYAL SCOT looking in a less than satisfactory state at Crewe, although it still had over two years more in service before withdrawal. The original Fowler-designed No. 6100 was the first of the class, built at the Queen's Park Works of the North British Locomotive Company, Glasgow, and delivered to the LMS on 14th August 1927. The class were an immediate success. Fifty were built by November of the same year and another twenty ordered, this time to be constructed at Derby Locomotive Works. In 1933 ROYAL SCOT was chosen to attend the 'Century of Progress' exhibition and toured the USA and Canada along with a train of LMS coaches, clocking up a total of 11,194 miles. As is well documented, for this trip it is believed that No. 6100 swapped identities with Derby-built No. 6152 THE KING'S DRAGOON GUARDSMAN, and this is evidenced by detail differences such as having a full set of Stanier pattern wheels and a Stanier type bogie. The engine was named ROYAL SCOT at the suggestion of Sir Henry Fowler after the London Euston to Glasgow Central express service, and not the Royal Scots Regiment as some believed. ROYAL SCOT was rebuilt by Sir William Stanier with a 2A taper boiler in June 1950 and it remained in service, allocated to Nottingham shed (16A), until withdrawal in October 1962 along with No. 6139, therefore becoming the first two of the class to be condemned. RCR5593

THE R.C. RILEY COLOUR COLLECTION · MIDLAND REGION

5TH MARCH 1961 • Ex-Midland Railway Class 1F 0-6-0T No. 41734 carrying out shunting operations at Staveley Iron & Chemical Co's works in Derbyshire. This engine was one of the last five of the type and had been used under an agreement of 1866 between the iron company and the Midland Railway for them to supply locomotives to shunt the works sidings for 100 years. 41734 was one of the 137 locomotives built by Mr. S. W. Johnson between 1878 and 1892, known as the 1377 class. RCR6051

5TH MARCH 1961 • Ex-Midland Railway Class 0F 0-4-0T No. 41533 pictured in between duties at Staveley Ironworks. Designed by Richard Deeley and built at Derby Works the loco dates from 1883 and was one of a class of ten, five built from new and the remaining five being converted from other locomotives. In British Railways days, 41533 was primarily allocated to Staveley (Barrow Hill) shed apart from short spells at Burton and Gloucester (Barnwood), clocking up 84 years service, being withdrawn on 25th December 1966.

Founded in 1863, the Staveley Coal and Iron Company Limited was an industrial company based in Staveley, near Chesterfield. The company exploited local ironstone quarried from land owned by the Duke of Devonshire on the outskirts of the village. It developed into coal mining, owning several collieries and also into chemical production, first from those available from coal tar distillation, later to cover a wide and diverse range. Part of the plant at Staveley was a sulphuric acid manufacturing unit. It was during World War I that the company developed its chemical operations beyond coal-tar chemicals and began production of sulphuric and nitric acids. During the war they also made picric acid, TNT and guncotton. Staveley Coal and Iron Company were the first company in Britain to manufacture sodium chlorate, the plant becoming operational in 1938 and they became the main competitor to Imperial Chemical Industries (ICI) in caustic soda production.

In 1960, the Staveley Iron and Chemical Company, taken over by Stewarts & Lloyds Limited, merged with the Ilkeston-based Stanton Iron Works to form Stanton and Staveley Ltd. In 1967, Stewarts and Lloyds became part of the nationalised British Steel Corporation; Stanton and Staveley were also incorporated. RCR6046

5TH MARCH 1961 • Johnson Class 1F 0-6-0T No. 41739 shunting Staveley Ironworks with a party of enthusiasts in attendance. RCR6056

RIGHT: 10TH MAY 1959 • Class 2P 4-4-0 No. 40542 at Millhouses shed. A 1913 Fowler rebuild of the Johnson Midland Railway designed locomotive, it was a Nottingham (Midland) based loco for most of its near 60 year working life.

Millhouses shed was located near Sheffield. Built by the Midland Railway in 1901 as Ecclesall engine shed, it was used mainly to stable passenger and mixed-traffic locomotives for use on trains from the nearby Sheffield Midland station. The shed was built next to Millhouses railway station; it had eight roads, and could handle about forty steam locomotives. Ecclesall shed was renamed Millhouses in 1920. Millhouses bore the shed code 25A, then 19B in 1935 and finally 41C in 1958.

During LMS ownership the shed was home mostly to 4-4-0 engines until the arrival of 'Jubilees' in 1937. In 1959 the shed had an allocation of 33 engines, including twelve of the aforementioned 'Jubilees'. 1960 saw the arrival of seven Royal Scots.

Closing on 1st January 1962, the shed's remaining engines were transferred to Canklow, Barrow Hill and Staveley (Great Central). After many years of standing derelict, in December 2015 the shed was demolished. RCR5716

9TH MAY 1964 • Two views of a very rare visitor to the former Great Central main line in the shape of Stanier 'Duchess' No. 46251 CITY OF NOTTINGHAM which hauled the first and final legs of an R.C.T.S. (East Midlands Branch) rail tour titled 'The East Midlander No. 7'. The train was made up of twelve coaches, being so popular that bookings had quickly exceeded the planned eight coaches. Above, the train departs from Leicester Central and (right) approaches Leicester South Passenger signal box.

As well as the use of the Pacific on routes not normally used by them, the passengers were able to tour both Eastleigh Works and Swindon Works. A delay to the tour was caused at Didcot due to photographers standing in the adjacent '4 foot' in front of an oncoming train – they were almost hit by a 'Hymek' diesel travelling in the opposite direction. RCR5569/RCR5568

THE R.C. RILEY COLOUR COLLECTION • MIDLAND REGION

THE EAST MIDLANDER NO. 7 RAIL TOUR – 1X36 – 9th May 1964	
Railway Correspondence and Travel Society (East Midlands Branch)	
46251	Nottingham Victoria – Loughborough Central – Leicester Central – Lutterworth – Rugby Central – Woodford Halse – Culworth Junction – Banbury Junction – Banbury General – Oxford – Didcot North Junction – Didcot
34038	Didcot – Newbury – Winchester Chesil – Eastleigh
30071	Eastleigh station – Eastleigh Works Yard
34038	Eastleigh Works Yard – Eastleigh station – Romsey – Salisbury – Westbury – Chippenham – Swindon Works Junction
46251	Swindon Works Junction – Swindon – Foxhall Junction – Didcot North Junction – (reverse of outward route) – Nottingham Victoria

18TH JUNE 1963 • Stanier Black Five 4-6-0 No. 44830 departs Leicester Central, and is about to pass Leicester North Passenger signal box and travel through the north of the city. Here the line was predominantly built on a lengthy blue brick arched viaduct that crossed over the River Soar and continued on an embankment towards Belgrave and Birstall station.

The ex-GCR line came under the control of the London Midland Region in 1958 and was generally regarded as an unnecessary duplication of existing north-south routes; as a consequence services were gradually run down. Most services were now in the hands of Black Fives, also in a run down state. The engine shed at Leicester Central closed in 1964, and freight services were withdrawn in June 1965. On 3rd September 1966 the line ceased to be a trunk route with the withdrawal of services to Sheffield and Marylebone, leaving Leicester Central operating a sparse DMU local service to Nottingham and Rugby; its closure came on 5th May 1969. **RCR5619**

ABOVE: 13TH MAY 1965 • Stanier 5MT 4-6-0 No. 44836 on a Down passenger express passing through Woodford Halse station. RCR5613

The station opened under the name Woodford and Hinton and served the nearby villages of Woodford Halse to the east and Hinton to the west, both in Northamptonshire. The station was renamed Woodford Halse on 1st November 1948. The village of Woodford Halse became notable for the role it played as an important railway centre. Originally it had seemed the railway would by-pass Woodford, as there were stations at Byfield (two miles west), and Moreton Pinkney (three miles south-east) which were both on the East and West Junction Railway (which became part of the S&MJR) and no other lines seemed likely to be built in such a thinly populated area. However, in the late 1890s the village found itself on the Great Central Railway's London Extension.

The station was a variation on the standard island platform design typical of the London Extension; here the less common 'embankment' type design was utilised, being reached from a roadway that passed beneath the line. It differed from the usual design in that it served what was effectively a four-way junction, and was therefore provided with a more extensive range of platform buildings and facilities. Situated near Woodford Halse North Curve Junction that linked the GCR with the S&MJR route between Stratford-upon-Avon and Towcester, a separate platform was provided on the west side for these destinations. Originally it was a timber structure but was later replaced by a concrete slab construction (as in the picture above), staff still referred to it as the 'wooden platform' though. Further south however was the more important Culworth Junction, a divergence point for a stretch of line 8¼ miles in length linking the GCR with the GWR at Banbury, enabling extensive and varied cross-country workings to take place.

North of the station was the locomotive depot, wagon and sheet repair shops, plus extensive marshalling yards, facilities that were originally intended to be located at Brackley. Local opposition forced the GCR to change its plans and the site moved to Woodford Halse. Located on top of a vast embankment, the site covered around 35 acres formed mainly from spoil taken from Catesby Tunnel. 136 terraced dwellings to house the railway workers were built on the east side of the embankment, together with a street of shops. This gave a small rural village an industrial look and increased the population to around 2,000. The depot and yards were a hive of activity, but not busy enough to ensure survival. On 5th April 1965 the marshalling yards closed, and on 5th September 1966, most of the GCR was closed completely, including all remaining lines converging on Woodford Halse.

THE R.C. RILEY COLOUR COLLECTION ♦ MIDLAND REGION

31st May 1958 • Fowler Class 4F 0-6-0 No. 44085 trundles through Stamford Town station on a mixed freight. RCR5801

Built by the Syston & Peterborough Railway – a subsidiary of the Midland Railway – the station opened in 1848. Built in a classic mock Tudor style, designed by Sancton Wood, it included a bell tower (seen above the Home signal) with a gilded weather vane bearing the initials SPR.

On the main platform side, the canopy on its cast-iron columns dates from the 1870s as does the wooden waiting shelter and canopy on the opposite side. The lattice footbridge which can be seen between the canopies is a standard Midland Railway pattern.

'Town' was added to the station name in 1950 to differentiate it from Stamford East station. East station closed in 1957, resulting in the Stamford to Essendine services being diverted to Town station until these services ceased in 1959. In April 1966 the station name reverted to Stamford when the Town suffix was dropped.

Midland Railway fire buckets and notice at Stamford Town station. RCR0044

ABOVE: 31ST MAY 1958 • Class 3P 4-4-2T No. 41975 approaches the bay platform at Stamford Town to form a service to Seaton Junction. The loco was introduced in 1923, being a Midland Railway and LMS development of Whitelegg's London, Tilbury & Southend Class 79.

In addition to those constructed by the LTSR and MR, 35 were delivered to the LMS – ten in 1923, five in 1925, ten in 1927, and a final 10 in 1930. The ten delivered in 1923 were to an outstanding order placed by the MR, the remainder were ordered by the LMS. The five delivered in 1925 were built by Nasmyth, Wilson and Company, with the other thirty built at Derby Works. In 1947 the LMS assigned them the numbers 1928–1975, to clear their previous numbers for new LMS Fairburn 2-6-4T locomotives, but none of these were applied before nationalisation in 1948, leaving British Railways to allocate the numbers 41928–41975. RCR5875

RIGHT: 12TH SEPTEMBER 1970 • Oakham Level Crossing signal box is a Grade II listed signal box based in the heart of Oakham on the west side of the tracks by the level crossing. Built in 1899 according to the Midland Railway design type 2b, Oakham's signal box is not only a recognisable local landmark, but was the prototype for an Airfix plastic model kit, making it famous nationally among railway and model enthusiasts. RCR2005

19TH MAY 1957 ◆ Stanier Black Five No. 45154 LANARKSHIRE YEOMANRY at Spital Bridge shed, Peterborough. Construction of the shed commenced in 1848 for joint use by the Eastern Counties Railway (ECR) and the Midland Railway. The facility included the shed itself, which was brick built with six tracks, plus water tanks able to hold 72,000 gallons. In addition, there were offices, stores, a coaling stage, turntable (originally 42ft but replaced by a 50ft version in 1899), water crane and workshops for locomotive repairs.

The Midland Railway line to Peterborough was constructed to thwart the embryonic Great Northern Railway and the MR built its own shed at Spital Bridge in 1872; its importance was due in great part to traffic off the Midland and Great Northern line that came in from Norfolk. RCR5677

THE R.C. RILEY COLOUR COLLECTION • MIDLAND REGION

30TH MAY 1959 • Fowler Class 2P 4-4-0 No. 40504 pilots Stanier Black Five 4-6-0 No. 44861 as they pass Kettering signal box with an Up express to St. Pancras. RCR5723

THE R.C. RILEY COLOUR COLLECTION ♦ MIDLAND REGION

14TH JUNE 1963 ♦ Nuneaton based ex-LMS 'Jubilee' 4-6-0 No. 45643 RODNEY passes through Narborough station (Leicestershire) after exiting the goods sidings. The picture was taken from the station footbridge in the direction of Leicester. Closed in March 1968, the station became the first in the country to reopen after the Beeching closures, this happening on 5th January 1970 after much lobbying by the local council, setting the trend for many more throughout the country. The goods shed is still extant (out of shot to the left of the train) and is now occupied by a builders' merchant, although all of the sidings that accompanied it are gone with houses built on the land they used to occupy. The line is still a busy one and is the main route from Leicester via Nuneaton to Birmingham New Street. An hourly passenger service calls at the station and it sees frequent container trains from the east coast to the distribution centres of the West Midlands. RCR5661

17TH JUNE 1963 • Stanier 8F 2-8-0 No. 48065 passes through Blaby station in Leicestershire. Situated on the Birmingham to Peterborough Line the station was opened in 1864 by the South Leicestershire Railway, which was taken over by the London and North Western Railway in 1867. The station was closed to passenger traffic in 1968.

In July 1914, local suffragettes Ellen Sheriff and Elizabeth Frisby, along with experienced arsonist Kitty Marion, armed with wood-shavings dipped in creosol (and an axe, to break in) trekked across a field in the middle of the night and burned the station down, causing £500 worth of damage. A campaign to reopen the station was launched in 2008 and a preserved Bagnall fireless steam locomotive (No. 2370) is being used to publicise the reopening campaign; it sits in a field close to the station site. RCR5769

4TH MAY 1962 • Johnson Class 2F 0-6-0 No. 58143 resting between shunting duties at Leicester West Bridge. RCR5813

The Leicester and Swannington Railway (L&SR) was one of England's first railways, being opened on 17th July 1832 to bring coal from collieries in north-west Leicestershire to Leicester. The construction of the railway was a pivotal moment in the transport history of the East Midlands, which was characterised by fierce rivalry between the coal companies of Leicestershire and Nottinghamshire. Through the latter half of the 18th century, the Leicestershire miners, using horses and carts, had been at a disadvantage compared to those in Nottinghamshire, who had access to the Erewash Canal and the Soar Navigation; in 1794 the latter was extended to Leicester. In 1828 William Stenson observed the success of the Stockton and Darlington Railway and, with John Ellis, and his son Robert, travelled to see George Stephenson when he was building the Liverpool and Manchester Railway. Stephenson visited Leicester on their invitation and agreed to become involved. The first meeting to discuss the line was held at the Bell Hotel in Leicester on Thursday 12th February 1829, where subscriptions amounting to £58,250 were raised. The remainder of the £90,000 was raised through Stephenson's financial contacts in Liverpool. The line obtained the Royal Assent on 29th May 1830 and the first part opened in 1832. A second Act on 10th June 1833 authorised two branch lines, the first was to Snibston Colliery and the second across the River Soar in Leicester to the Soar Land and Pringle Wharfs.

The line was ceremonially opened on 17th July 1832 by a special train for the L&SR directors and 300 guests. The train was hauled by the locomotive COMET and the following day public services began, making it the first steam locomotive hauled railway in the Midlands. The branch opened throughout for passenger traffic on 27th April 1833. In an Act dated 2nd July 1846 the L&SR was dissolved and vested into the Midland Railway, this taking effect from 1st January 1847 and making it the oldest component of the Midland Railway. A Midland Railway Act on 2nd July 1847 authorised the MR to abandon the previously authorised extension to the west from Coalville to Burton-upon-Trent (authorised in 1846) and to propose a new route for the same extension. This Act also authorised the doubling of the line from Coalville to Bagworth and Thornton to Desford, as well as a deviation to remove the old Bagworth incline. The final amendment under this Act was a new route into Leicester from Desford to Knighton Junction which fully opened to traffic on 1st August 1849.

The original line from Desford Junction to Leicester via Glenfield then became known as the 'West Bridge Branch', passenger traffic continuing on this section until 24th September 1928.

2ND MAY 1963 • One of the Johnson Class 2F 0-6-0s, No. 58148 from Coalville shed, working a goods train on the West Bridge, Leicester to Desford Junction leaving Glenfield Tunnel. For many years engines of this class were retained to work the line as they were sufficiently small to pass through the narrow bore of the 1,796 yard long Glenfield tunnel, visible in the background. The Class 2Fs reigned until December 1963 when they were replaced by slightly modified Standard Class 2 2-6-0s. No. 58148 was a member of the Midland 1142 class built in 1875-76 by contractors. During the course of their long lives most of them were rebuilt with G6 boilers, new frames and Deeley cabs.

Glenfield Tunnel, when opened by the Leicester & Swannington Railway in 1832, was one of the longest steam railway tunnels in the world at 1 mile and 36 yards; it was designed by the famous railway engineer George Stephenson and built between 1830 and 1832 under the supervision of his son Robert. The biggest obstacle to this project was a ridge extending from Gilroes to Glenfield village that required a tunnel. The project ran heavily over budget but resulted in a tunnel that remained in use for 130 years. The project to build this tunnel was very testing for its engineers, involving techniques that were then virtually untried. Faulty trial drillings suggested the bore would be through stone and clay, when, in fact, much of the bore would turn out to be in running sand. This necessitated a great deal more work and expense. The tunnel had to be lined throughout in brickwork that was between 14 and 18 inches thick, backed by a 'wooden shell' where running sand was encountered. Bricks for the lining were made in an on-site kiln. Owing to the problems encountered, the tunnel construction ran well over the proposed budget of £10,000, finally costing £17,326 12s 2½d. RCR5824

THE R.C. RILEY COLOUR COLLECTION • MIDLAND REGION

4TH MAY 1962 • Johnson Class 2F 0-6-0 No. 58143 crosses over Fosse Road bridge with a lengthy freight after departing Fosse Road Wharf, part of the Leicester West Bridge goods yard. A wonderful everyday scene captured by Dick Riley showing the advertising of the day alongside on the bridge – they include Heinz Baked Beans, Hovis and Danish bacon – and the impromptu allotments are a resourceful use of land to 'grow your own produce'. RCR5827

4TH MAY 1962 • With smoke still filtering out of the narrow bore of Glenfield Tunnel, No. 58143 eases into the disused station. Note the rear of the train is still to see daylight, no doubt the guard was firmly enclosed in his van. Passenger services to this small station ceased on 24th September 1928 with goods traffic continuing until withdrawn on 6th December 1965, although trains continued from Groby Granite sidings until September 1967. Ammunition sidings were located at Glenfield during the First World War. RCR5831

THE R.C. RILEY COLOUR COLLECTION • MIDLAND REGION

ABOVE: 3RD MAY 1963 • No. 58148 engaged in shunting duties at Glenfield, including dropping supplies off at the adjacent Glenfield Coal & Coke Co. There was a sharply-curved private siding into the merchants' small yard but by the time the photos were taken it had been removed. RCR5842

RIGHT: 3RD MAY 1963 • The old Glenfield station building of 1832. Built in the style of a bow-fronted toll house, it combined as the crossing-keeper's house and booking office. RCR1802

ABOVE: 3RD MAY 1963 • No. 58148 at Groby Sidings where the junction connected with The Groby Granite Company's private branch. Midland Railway property from the junction extended for 5 chains (110 yards) where the private line became single track. The line in between the locomotive and wagons led to the former Glenfield Premier Brick & Terra Cotta Works. RCR5834

BELOW: 5TH MAY 1962 • Moving down the line we see a train headed by No. 58143 arriving at a dilapidated Ratby station. A siding first appeared here in 1850 while the station was completed in 1876 with the platform being extended in 1887. The public house to the left of the engine issued tickets before the station buildings were completed. The pub is still open and is appropriately called 'The Railway Inn'. The goods yard at Ratby, which was located to the right of the train, closed on 4th October 1954 and the sidings were lifted in 1959, and as at Glenfield, passenger services ceased on 24th September 1928. RCR5814

ABOVE: 5TH MAY 1962 • No. 58143 shunting at Desford Junction. The Leicester & Swannington was officially taken over by the Midland Railway on 27th July 1846 who then built a through route from Knighton Junction (a couple of miles south of Leicester) to Burton-upon-Trent. This utilised the Desford to Coalville section of the former Leicester & Swannington, thereby creating Desford Junction which connected the line to Ratby, Glenfield and Leicester West Bridge.

The junction opened on 1st August 1849. Two signal boxes had been provided at Desford Junction in 1876; these were subsequently replaced by one in 1882. This signal box was then replaced in July 1917 by the one pictured above which had 40 levers. It was unusual in that the name was displayed on the box and repeated in the 'V' of the junction (located above the oil drum). RCR5821

3RD MAY 1963 • A view of Desford station. The 16-lever Midland Railway signal box dates from 1897, while the station buildings and house were built in 1848. A shelter on the Up platform was added in 1862. Of note is the original platform which measured only 10 inches above the rail height and remained in its original state because of the obstructing doorways, until closure in September 1964. RCR1905

ABOVE: 3RD MAY 1963 • Stanier 8F 2-8-0 No. 48619 moves off shed (out of shot to the right) at Coalville with Johnson 2F No. 58143 standing opposite.

A three road brick-built shed was completed here by the Midland Railway in 1890 along with a standard coal stage and 50ft turntable. Around this time the number of engines required at Coalville was steadily increasing due to the amount of coal traffic handled from the surrounding collieries. Soon after opening it was found that the water supply was inadequate and in 1893 a new well was completed at a cost of £800. It was given the code 10B as it was a sub-shed of Leicester (10).

Under the LMS in 1935 the shed was elevated in status and control was transferred to Derby, with Coalville becoming 17C. Various improvements followed with the ashpit being renewed in 1939, a further extension in 1944 saw the deep well receive a new pump – softening apparatus for the supply was also provided.

More modern locomotives in the shape of 8Fs arrived during the Second World War, staying until traffic declined. The shed changed identity again in 1958, becoming 15D under Wellingborough and changing for the last time in 1963 to 15E until closure in October 1965. RCR5765

ABOVE RIGHT: 28TH SEPTEMBER 1957 • No. 47236 stands by the coaling stage at Horninglow shed in Burton-on-Trent. The town, just inside the Staffordshire border, was deep in the heart of Midland territory, but this shed was owned by the LNWR, who successfully made a territorial invasion into the other LMS constituent company's area. The shed was closed in September 1960. Note the substantial malthouses, now demolished, but evidence of Burton's importance then as one of the major brewery towns in England. RCR5899

BELOW RIGHT: 26TH MAY 1959 • Fowler class 4F 0-6-0 No. 44420 pictured exiting the exchange sidings at Dixie Yard, Burton-upon-Trent. Introduced by the LMS in 1924, this is one of a total of 580 locos that were a post-grouping development of the Midland Railway Class 3835 introduced by Fowler in 1911. RCR5810

Burton's industrial railway system extended over an area of four square miles, this comprising almost 90 miles of track with 32 level crossings. The largest of the breweries – Bass – operated 16 miles as a private branch network and by 1925 accounted for an average freight movement of 1,000 wagons daily. When these pictures were taken in the late 1950s, there was still a large volume of traffic and steam locos would be seen in the town until the mid- to late-1960s before brewery traffic transferred to road transport.

THE R.C. RILEY COLOUR COLLECTION • MIDLAND REGION

THE R.C. RILEY COLOUR COLLECTION • MIDLAND REGION

ABOVE: 12TH APRIL 1958 • Ex-LNWR Beames Class G2 0-8-0 No. 49410 stands next to the coaling stage at Burton-on-Trent Horninglow shed. Allocated to Stafford shed, the commendably clean 7F, dating from January 1922, would have a further 18 months in service before withdrawal.

The largest engine shed in Burton-on-Trent belonged to the Midland Railway. The LNWR shed, a smaller affair known as Horninglow, came under the control of the MR shed shortly after Grouping in 1923. A third shed, that of the North Staffordshire Railway, was closed and demolished.

Horninglow, as a sub-shed, was recoded 17B in the 1935 reorganisation, retaining this code until closure on 12th September 1960. RCR5736

LEFT: 13TH APRIL 1958 • The Midland Railway notice leaves the drivers in no doubt as to what to do, whether this sign is at Burton or Horninglow shed is not recorded. RCR2103

THE R.C. RILEY COLOUR COLLECTION • MIDLAND REGION

12TH APRIL 1958 • Work weary ex-MR Class 3F 0-6-0 No. 43584 waits patiently by a signal gantry at Burton-on-Trent. The class was designed by Samuel Johnson and introduced to service in 1885, they were then rebuilt from 1916 by Henry Fowler with Belpaire boilers. RCR5808

24TH MAY 1959 • Trent, near Derby, was a station surrounded by fields and without main road access. It was intriguing for the timetable connoisseur as two trains departing from both sides of the island platform might both be bound for St. Pancras. The Midland Railway built a series of complex junctions at Trent to allow access to the Erewash Valley from all directions and from there the main line forged south through Loughborough and Leicester towards London. Trent station was built in 1862 as an interchange and on peak days it handled more trains than both Nottingham Midland and Victoria combined. As the Beeching cuts of the 1960s bit deeply into railway services, it was on 1st January 1968 at 00.01 that the final train departed from the station. The fine array of Midland semaphore signals surrounding Trent Station North Junction signal box is evident in this photo as a mineral train passes a 'Peak' on a passenger working. RCR2041

24TH MAY 1959 • Jubilee No. 45565 VICTORIA calls at Trent station with the 11.18am Bradford–St. Pancras. RCR5642

THE R.C. RILEY COLOUR COLLECTION • MIDLAND REGION

STEPHENSON LOCOMOTIVE SOCIETY (Visit to Derby and Nottingham) – M855 – 27th September 1959
Hauled throughout by Midland Compound No. 1000
Birmingham New Street – Aston – Lichfield Trent Valley – Wichnor Junction – Burton-on-Trent – Stenson Junction – Derby
Derby – Melbourne Junction – Chellaston Junction – Sheet Stores Junction – Trent – Stapleford & Sandiacre
Stapleford & Sandiacre – Meadow Arrival
Meadow Arrival – Stapleford & Sandiacre
Stapleford & Sandiacre – North Erewash Junction – Long Eaton Junction – Nottingham Midland
Nottingham Midland – Trent – Sheet Stores Junction – Chellaston Junction – Stenson Junction – Burton-on-Trent – Wichnor Junction – Tamworth High Level – Kingsbury – Water Orton – Saltley – Birmingham New Street

ABOVE: 27TH SEPTEMBER 1959 ♦ Johnson Midland Compound 4-4-0 No. 1000, looking resplendent in its original Midland Railway livery, draws the attention of young enthusiasts and a policeman at Nottingham Midland. The engine crew have a chat prior to working the final leg of the SLS Special (M855) back to its starting point of Birmingham New Street. During the tour visits were made to Derby, Toton (right) and Nottingham depots.
RCR6120/RCR6124

THE R.C. RILEY COLOUR COLLECTION ♦ MIDLAND REGION

THE R.C. RILEY COLOUR COLLECTION • MIDLAND REGION

27TH SEPTEMBER 1959 • Johnson Midland Compound 4-4-0 No. 1000 passes under the fine array of signal gantries controlled by Derby Station North box on its SLS Special tour described on the previous page. **RCR6132**

THE R.C. RILEY COLOUR COLLECTION • MIDLAND REGION

24TH MAY 1959 • Three-cylinder Compound Class 4P 4-4-0 No. 41157 pictured leaving Derby Midland station around the tight curve towards Spondon. RCR5722

25TH MAY 1959 • Ivatt Class 2MT 2-6-0 No. 46440 engaged in shunting duties with a distant view of the works from Derby Station North signal box. RCR5699

24TH MAY 1959 • Fowler Class 2P 4-4-0 No. 40553 leads a line of condemned locomotives at Derby alongside the main line to Birmingham. A Nottingham-allocated loco for all of its British Railways life, 40553 had been withdrawn in November 1958 and scrapping wouldn't take place until January 1960, this took place at Cashmore's (Great Bridge). RCR5708

24th May 1959 • The signalman enjoys the Spring sunshine as he watches Stanier Class 6P 'Jubilee' 4-6-0 No. 45662 KEMPENFELT pass under one of the many splendid gantries in the area after departing Derby Midland station with a Down express.

The immediate area around Derby station had numerous signal boxes, each controlling a small area or a limited number of running lines. At its height Derby station was controlled by six signal boxes and they were all located within its immediate confines. They were: London Road Junction, Derby Station 'A', Derby Station 'B', Engine Sidings No. 1, Engine Sidings No. 2 and Derby Junction.

The signal box featured here is Derby Junction which was situated on the east (Up) side of the lines to the south of the triangle that led to Chaddesden sidings. It was a Midland Railway Type 2b box comprising 52 levers which opened in October 1892, remaining in use until July 1969. RCR5657

THE R.C. RILEY COLOUR COLLECTION • MIDLAND REGION

THE R.C. RILEY COLOUR COLLECTION • MIDLAND REGION

25TH SEPTEMBER 1955 • Looking in considerably better condition than in the photo on pages 68-69, rebuilt 'Royal Scot' Class 7P 4-6-0 No. 46100 ROYAL SCOT is pictured near to the coaling stage at Derby after replenishing its tender. RCR5600

THE R.C. RILEY COLOUR COLLECTION • MIDLAND REGION

24TH MAY 1959 • Stanier 'Jubilee' 4-6-0 No. 45585 HYDERABAD approaching Derby on a Down express service. Viewed from Way & Works box, the train is passing the sidings of the same name on the right, with Etches Park carriage shed on the left and the Derby Coalgas & Coke works in the left background (where Pride Park now stands). Derby No. 4 shed is just out of sight to the left. The Railway Technical Centre would be established to the right of the train during the 1960s. RCR5648

12TH MAY 1964 • Stanier 8F 2-8-0 No. 48145 on a train of empty minerals near Riddings Junction after passing through Pye Bridge station, which was situated on the Erewash Valley line in Derbyshire. RCR5759

24TH SEPTEMBER 1961 ♦ Samuel Johnson Midland Railway Class 115 4-2-2 No. 118 pictured at Derby Works.

In the late Victorian era, the Midland Railway was noted for its single-driving wheel express locomotives, commonly known as 'Spinners'. No. 118 was completed at Derby Works in 1897, one of a class of fifteen locomotives. They were built in two batches between 1896 and 1899, No. 118 being part of the first batch of five.

The driving wheels had a diameter of 7ft 9½ins with the class being able to attain speeds of up to 90 mph.

Renumbered 673 by the Midland Railway in 1907, it retained this number when the LMS came into existence in 1923. It was withdrawn from service in 1928 and repainted in Midland Railway colours with its original number, but no longer in working order. It is the oldest preserved engine in Britain to have been built new with piston valves.

Stored in Derby works for many years, it was periodically brought out for special occasions. As No. 673 it was restored to working order so that it could take part in the Rainhill Trials 150th Anniversary cavalcade that took place between 24th-26th May 1980. It is now a static exhibit at the National Railway Museum in York. RCR6155

THE R.C. RILEY COLOUR COLLECTION • MIDLAND REGION

18TH MAY 1955 ◆ A photo taken in the beautiful Peak District of an unidentified Stanier 8F on an Up freight crossing Monsal Dale Viaduct (also known as Headstone Viaduct). In 1970, a Grade II listing was placed on the viaduct and since 1980 this section of line has formed part of the Monsal Trail, an 8½ mile (13.7km) walk and cycleway formed on the Midland Railway's former Manchester to St. Pancras line. RCR5749

THE R.C. RILEY COLOUR COLLECTION • MIDLAND REGION

The Cromford and High Peak Railway (C&HPR) ran between the Cromford Canal wharf at High Peak Junction and the Peak Forest Canal at Whaley Bridge. Completed in 1831, it was built to carry minerals and goods through the hilly and rural terrain of the Peak District within Derbyshire. The route was marked by a number of rope worked inclines.

Sheep Pasture Incline, as seen in the image above, was one of six gradients on the line. The incline was opened in 1830 and was in use until 1967 when the line from Whaley Bridge to Middleton was closed, with the section to Cromford closing in 1963. Today, the incline is used as part of the High Peak Trail. Sheep Pasture Incline was on a 1 in 8 gradient, similar to others on the railway. Due to falling traffic, the entire railway had closed by 1967.

ABOVE: 9TH OCTOBER 1959 • One of ten small Class 0F 0-4-0STs, No. 47007 is seen working at Sheep Pasture, Derbyshire. Five locomotives were built by Kitson's to Stanier's requirements for the LMS in 1932, while a further five were included in the 1953-54 BR construction programme and it fell to Horwich Works to build them.

47007 was outshopped in October 1953 and put to work initially at Birkenhead Mollington Street shed (6C). After a short working life of just over 10 years it was withdrawn in December 1963. RCR6140

RIGHT: 18TH MAY 1966 • A view of the foot of the incline at High Peak Junction, where the railway meets with the Cromford Canal. A Class 03 diesel shunter is just visible. The C&HPR workshops were located here. RCR1789

29TH AUGUST 1959 • One of Fowler's Class 5XP, later 6P, Patriot class express passenger engines makes a fine picture standing at Willesden shed in the low sunlight. Portrayed is No. 45511 ISLE OF MAN, built at Crewe in 1932 but not named until 1938. The crest of the island above the nameplate is clearly visible. The Patriots were built to provide a passenger engine with a wider route availability than the recently introduced Royal Scots and which might do the work of the existing Claughtons without incurring the same heavy maintenance and fuel costs of the older engines. Even the rebuilt large-boilered Claughtons had not resolved this problem, so Sir Henry Fowler decided to put the 3-cylinder chassis of the Royal Scot class under the enlarged Claughton boiler. The result was the two prototypes, classified 5XP, nominally rebuilt Claughtons, but in reality very little was retained. They proved superior in performance to the enlarged Claughtons, with repair costs averaging 50% less than the older engines. Fifty more Patriots were built between 1932-34 and proved more than satisfactory in service. No. 45511 amassed almost 1.3 million miles before withdrawal from Carlisle Upperby (12B) in February 1961. **RCR5673**

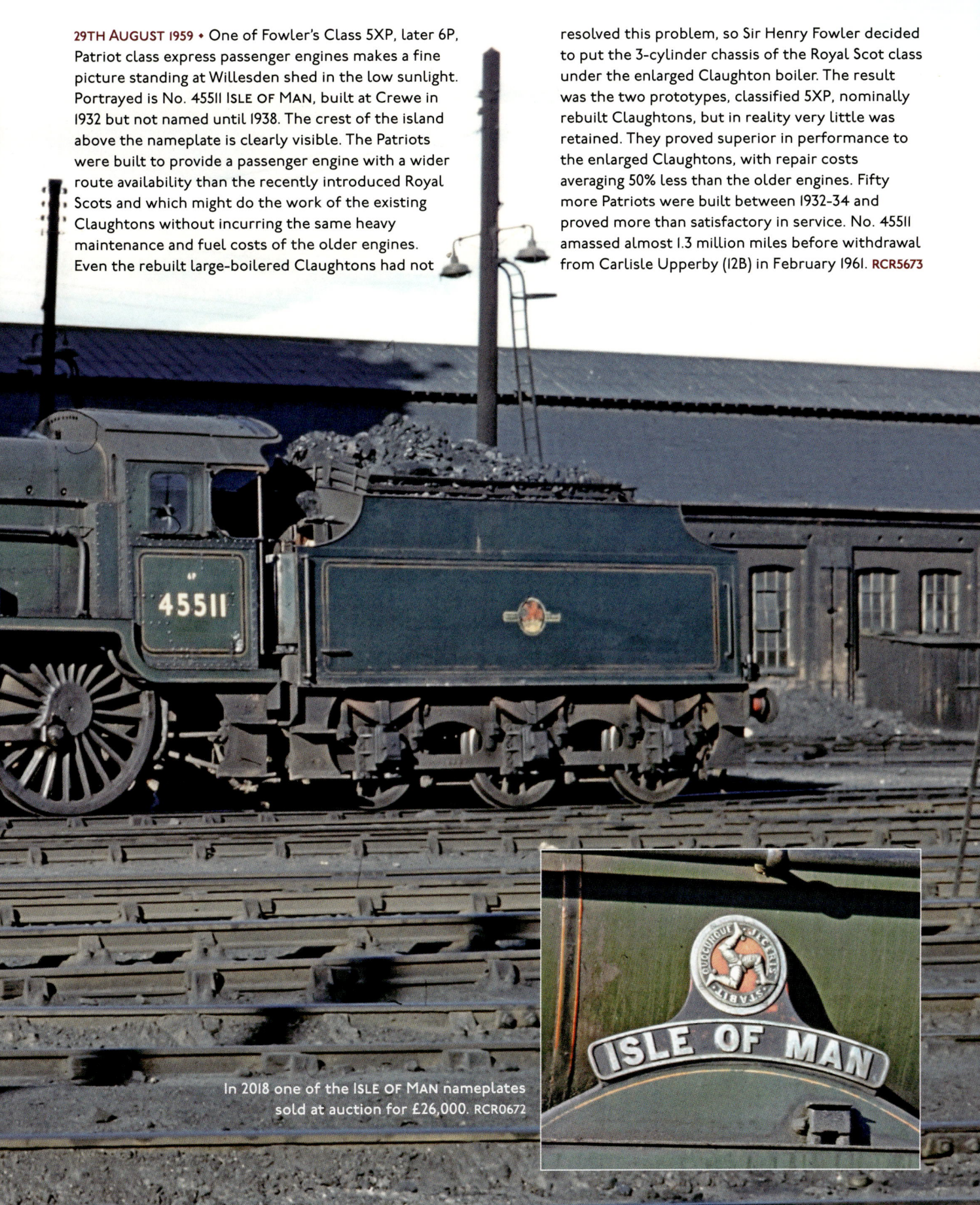

In 2018 one of the ISLE OF MAN nameplates sold at auction for £26,000. RCR0672

THE R.C. RILEY COLOUR COLLECTION • MIDLAND REGION

18TH MARCH 1961 • First of the 'Jubilee' class, No. 45552 SILVER JUBILEE on shed at Willesden. RCR5664

As with many of the early railway companies with terminus stations in the capital, the London and North Western Railway found it necessary to construct additional freight facilities as close to London as possible. Available land was purchased at Willesden where large carriage sheds, extensive freight yards, a wagon repair shop and gas works were constructed as well as a suburban station, Willesden Junction. An engine shed was built to house the freight and shunting locos using the new yards, this relieving Camden shed of having to provide engines for this purpose. Work on a new engine shed commenced in the early months of 1873; opening later the same year it could accommodate 48 locomotives on its 12 roads, covered by three hipped roofs over four tracks each.

In 1898 Webb enlarged the shed by lengthening it so that the capacity was increased to 60 engines, becoming one of the most important depots on the LNWR. It became responsible for approximately 150 engines by 1912; and included a reasonably sized repair shop.

From an early date the shed had been equipped with water treatment apparatus and was a favoured depot for modern aids. A mechanical coaler was built in 1920 at a cost of £7,000 followed a year later by an ash lifting plant and the provision of electric power to all of the shed equipment, taking the total spent to almost £10,000.

Throughout the 1920s the shed became busier and before the decade was out it required further accommodation to be constructed. In 1927 a contract was issued to build a modern roundhouse; this was to be located on spare ground to the east of the existing shed. The roundhouse was completed in October 1929, the facilities being more spacious and a significant improvement on the existing straight shed which was showing signs of decay.

Willesden, coded 1A from 1935, was always considered primarily as a freight locomotive shed. At this time a large proportion of its allocation consisted of 0-8-0 and 2-8-0s. This was maintained up to the 1950s and 'Crabs' for faster freights and 4F 0-6-0s for local work were also common occupants of the shed.

The shed had been one of the first to operate diesel shunters and by the mid-50s over 30 were in use, assisted by class 3F tanks, working the vast amount of sidings and yards in the area. The LMS diesels, Nos. 10000 and 10001, were based at Willesden when first introduced. As can be seen on these pages 'Patriots', 'Jubilees' and 'Royal Scots' were based there in later years for use on express passenger duties, supplemented by 2-6-2 and 2-6-4 tanks for suburban and empty stock duties.

The shed closed on 27th September 1965, being demolished and replaced by a new freightliner terminal.

THE R.C. RILEY COLOUR COLLECTION • MIDLAND REGION

ABOVE: 21ST SEPTEMBER 1955 • Rebuilt 'Patriot' No. 45529 STEPHENSON outside Camden shed. Previously a Camden machine, at the time of the photograph this engine was based at Crewe North where it remained until reallocating to Willesden in January 1961. It was destined to have a life of just under 31 years being withdrawn early in 1964. In the background is 'Britannia' No. 70048. RCR5647

BELOW: 11TH APRIL 1964 • Rebuilt 'Royal Scot' No. 46115 SCOTS GUARDSMAN with a full tender and ready for work. RCR5612

18TH MARCH 1961 • Fowler Class 3P 2-6-2Ts Nos. 40042 and 40049 stored at Willesden shed. Neither locomotive was officially withdrawn until 22nd July 1961, it would be another year before scrapping took place. RCR5859

THE R.C. RILEY COLOUR COLLECTION • MIDLAND REGION

18TH MARCH 1961 • Ivatt Class 2MT Mogul No. 46424, with its lined black livery showing up nicely in the sunshine, moves through the yard at its home shed of Willesden. RCR5703

8TH AUGUST 1965 • Bescot-allocated Stanier 8F 2-8-0 No. 48674 pictured at Willesden on the rear of an engineer's train. RCR5772

THE R.C. RILEY COLOUR COLLECTION • MIDLAND REGION

ABOVE: 29TH SEPTEMBER 1959 • 'Princess Coronation' Class 8P Pacific No. 46245 CITY OF LONDON on the Down 'Caledonian' at Northchurch. RCR5573

'The Caledonian' service was introduced in 1957; it was the post-war successor to the 'Coronation Scot' Euston to Glasgow express service, its only intermediate stop being at Carlisle. The Up train departed Glasgow at 8.30am, the Down train left Euston at 4.15pm. In 1958 additional 'Caledonian' services were introduced, departing Euston at 7.45am – known as 'The Morning Caledonian' – and Glasgow at 4.00pm labelled 'The Afternoon Caledonian', these were short-lived and had disappeared from the timetable in less than a year. 'The Caledonian' was gradually taken over by diesel power in the shape of English Electric Type 4s with extra stops added at Stafford, Crewe and Wigan. The service lasted until the end of the 1964 summer timetable and then withdrawn.

RIGHT: 21ST SEPTEMBER 1960 • An unidentified Stanier 'Black Five' emerges from one of the single bores of Northchurch Tunnel with an Up parcels train. The two single bores were for the Slow lines with the double bore (out of picture on the left) being for the Fast lines.
R.C. RILEY COLLECTION • RCR5641

Northchurch Tunnel is located at the western edge of Berkhamsted in Hertfordshire on the West Coast main line. It was originally constructed for the London and Birmingham Railway between 1836-37. This work was awarded to W. & L. Cubitt as part of contract 5B (Berkhamsted). The same company also won contracts 4B and 6B (Kings Langley and Aldbury) and were thus responsible for building a total length of 9¼ consecutive miles of railway. They also constructed the one mile Euston extension at the southern end of the line.

14TH MAY 1959 • Fowler class 3P 2-6-2T No. 40022 departs from Moorgate with the 5.11pm suburban service for City workers via the Widened Lines to St. Albans. It is alongside two Metropolitan 'T' stock EMU compartment trains in London Transport brown livery, one of which is bound for Watford.

No. 40022 was one of a batch of 20 locos (40021–40) of this class fitted with condensing gear for working around London. The class totalled 70, all of them being constructed at Derby Works between 1930 and 1932, No. 40022 was the last of the class to be withdrawn in December 1962.

Moorgate station dates from 1865 but increased traffic required the line from Kings Cross to be widened to four tracks the following year with completion in February 1868. Suburban services from the Midland Railway ran via Kentish Town and the Great Northern Railway ran via Kings Cross. Additional information for this photograph is that it was taken from scaffolding being used for development of property near to the station. RCR5861

22ND AUGUST 1959 • Stanier 'Black Five' 4-6-0 No. 45260 works a Margate to Derby service, having just passed through Kensington Olympia station. RCR5637

R.C.T.S. (London Branch) – SOUTHEND CENTENARY SPECIAL
11th March 1956 – Hauled throughout by No. 80

Bishopsgate Goods – Bethnal Green – Stratford – Forest Gate – Barking – Rainham – Tilbury Town – Pitsea – Southend Central – Shoeburyness

Shoeburyness – Southend Central – Pitsea – Laindon – Upminster – Barking – West Ham – Gas Factory Junction – London Fenchurch Street

Designed by Thomas Whitelegg, No. 80 was built in 1909 at Robert Stephenson and Company in Darlington and was delivered direct to the Imperial International Exhibition at Shepherd's Bush for which it was adorned in a special lavender livery and temporarily named SOUTHEND-ON-SEA. It was built to haul the heavier commuter passenger trains on the London, Tilbury and Southend Railway. Subsequently it carried the name THUNDERSLEY which is a district in the north of the Castle Point Borough in south-east Essex.

New boilers for the class were built at Crewe as late as 1948, and one was fitted to No. 80 in January 1949, after which it almost immediately entered storage at Wellingborough until February 1950, when it was transferred to Shoeburyness; this was its base until 1953 when it moved to Toton. Other engines of the class ended their days there as shunters until 1960, despite their great unsuitability.

The four locomotives ordered by the LTSR were numbered 79–82 and were named after places in Essex, near the LT&SR route. After absorption by the Midland Railway in 1912, they were renumbered 2176–2179 and their names were removed. The Midland gave them the power classification 3P, and later continued construction; an order for 10 locomotives was delivered in 1923, just after grouping. Following transfer to the Eastern Region in March 1956, No. 80 was withdrawn from service having completed 1,382,282 miles whilst in service. Although some LMS built engines survived in service until 1960, THUNDERSLEY was the last LT&SR engine to be withdrawn.

LEFT: 11TH MARCH 1956 • No. 80 THUNDERSLEY at Bishopsgate Goods prior to taking the special to Shoeburyness. RCR5876

BELOW: THUNDERSLEY gets the Southend Centenary Special away from Bethnal Green. Note the LT&SR 3rd class coach – No. 283 – behind the loco; it remained in the consist throughout the tour. RCR5888

15th December 1962 • LNWR 2-2-2 No. 3020 Cornwall at Nine Elms depot waiting to be transported to the Museum of British Transport in Clapham.

Cornwall was designed by Frances Trevithick in collaboration with Thomas Crampton and completed at the LNWR's Crewe works in 1847, numbered 173. It was completely reconstructed in 1858 to a 2-2-2 wheel arrangement and was renumbered to 3020 in 1885. Its present condition dates from 1887.

The locomotive has single 8 foot 6 inch driving wheels, which made it capable of high speeds. It worked until 1905, mainly hauling express trains between Liverpool and Manchester, Crewe and Euston. Between 1913 and 1922 it was fitted with a combined tender and saloon and used to haul the Chief Mechanical Engineer's Special Saloon.

In its original 4-2-2 form Cornwall was exhibited at the Great Exhibition of 1851. It took part in the centenary celebrations of the Stockton & Darlington Railway in 1925 and was finally withdrawn from traffic in 1927. It was then normally kept in the Crewe Works paint shop when not being displayed at exhibitions. The locomotive took part in the Liverpool & Manchester Centenary in 1930, which was the last time it was in steam.

In a letter to the National Railway Museum it has been claimed that the locomotive achieved a speed of 85-90mph in 1919. This followed the locomotive conveying Charles Bowen-Cooke to Euston where he realised that he had left important paperwork at Crewe and Cornwall was dispatched back to Crewe to collect the missing papers for the late afternoon.

In July 1920 No. 3020 took Charles Bowen-Cooke in the CME saloon to Euston to visit his doctor. Rather than returning light engine to Crewe, it was coupled in front of Claughton 4-6-0 No. 1914 Patriot at the head of the 'Midday Corridor'.

Cornwall was lent to the Severn Valley Railway in August 1979 by the Department of Education and Science, with the intention of being restored to working order. The locomotive was scheduled to take part in the Rocket 150 celebrations at Rainhill, but was forced to withdraw when a BR boiler inspector put his hammer right through the front ring of the boiler barrel. After further consideration, the cost of repairs to the boiler were deemed too expensive by the Severn Valley Railway and the Department of Education and Science.

No. 3020 Cornwall returned to York in September 1982 and is part of the National Collection. It has been exhibited at the Museum of British Transport at Clapham and the NRM at York and Shildon. In 2017 it was loaned to the Buckinghamshire Railway Centre, initially for a two-year period, where it currently resides. RCR6097

THE R.C. RILEY COLOUR COLLECTION

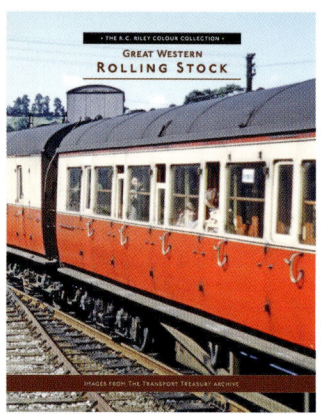

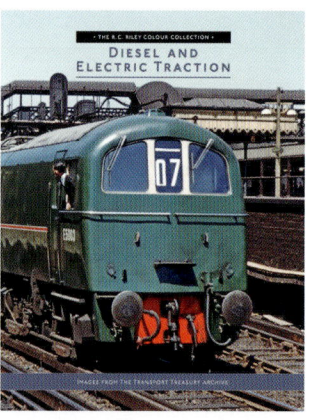

Transport Treasury Publishing are proud to present a unique series of 15 colour albums featuring the best of the R. C. Riley colour archive. To be released at intervals and printed in strictly limited numbers, the series will grow into a unique record both of the contemporary railway scene and also that of the work of one of Britain's leading transport photographers.

www.ttpublishing.co.uk

PUBLISHED BY TRANSPORT TREASURY PUBLISHING LTD.